# Modelling
# STIRLING
## and
# HOT AIR ENGINES

# Modelling STIRLING and HOT AIR ENGINES

## JAMES G. RIZZO

 **Patrick Stephens, Wellingborough**

*First published in 1985*

British Library Cataloguing in Publications Data

*Rizzo, James G.*
*Model Stirling and hot air engines.—*
*(PSL model engineering guide; 2)*
*1. Stirling engines—Models*
*I. Title*
*621.4      TJ765*

*ISBN 0-85059-736-6*

Patrick Stephens Limited is part of
Thorsons Publishing Group

*Photoset in 10 on 11 pt  Garamond by MJL Limited, Hitchin, Herts. Printed in Great Britain on 100gsm Fineblade coated cartridge, and bound, by Butler & Tanner, Frome, Somerset, for the publishers, Patrick Stephens Limited, Denington Estate, Wellingborough, Northants, NN8 2QD, England.*

# CONTENTS

# ACKNOWLEDGEMENTS

No engineering or technical book of this nature can be undertaken alone and this is no exception. The encouragement and assistance I have received have to be recorded somehow, not because these people expect it, but because it is a source of great personal satisfaction to me that so many people can be so helpful.

The person who started me writing about this subject certainly deserves very special mention: Mr Reno Borg, now the Director of Education in Malta, who by his encouragement in supporting my two previous publications, and his invitation to lecture to technical colleges on this subject, gave me the necessary push to write more about the hobby. Four persons were directly involved in the technical preparation of the book: Mr Gaetan Cilia, with his great technical drawing and designing experience; Mr Joe Camenzuli with his photographic expertise; Miss Lora Gauci who quite happily devoted a great deal of her spare time to type and retype the pages, and Mr Emanuel (Wally) Galea who cheerfully designed and constructed the two electronic projects in Chapter 16.

I sought and obtained invaluable assistance from my fellow model engineers, Mr Joe Simler, Professor Charles J. Camilleri, Mr Albert N. Debono and Mr Cristinu Vassallo who were prepared not only to advise on technical engineering problems but also to lend me their models to be included in the book; to these and other members of the Association of Model Engineers, Malta, I can only say 'thank you!'

My mentor, Mr David Urwick, formerly of Malta, now residing in England, was a great help when I started on this hobby; he spent many an hour advising and guiding me in my early days of 'Stirling' it, — he still is very helpful and most willingly sent me various photographs and papers of his engines. Other model engineers like Mr Andy Ross of Arizona, USA, Mr Robin Robbins of North Wales and Mr F. J. Collins of Surrey, were only too willing to help. Mr R. Robbins in particular advised on some of the early models. A model engineer who is no longer with us, Mr F. Brian Thomas, a surgeon, was also always ready to explain and help with problems; I am grateful to Mrs E. Thomas for allowing me to publish parts of her husband's article from *Model Engineer*.

Mr E. H. Cooke-Yarborough and Dr Colin D. West, formerly of the Atomic Energy Research Establishment have been a great source of information on two Harwell achievements, the thermo-mechanical generator and the fluidyne. Not only did they send me very interesting information but they readily undertook to discuss my problems on these two subjects. I also record with appreciation the help given to me by Philips of Eindhoven, Netherlands, and the editor of *The Engineer*, London.

To my family, who understood the need for my seclusion and many hours in the 'workshop', and especially to my eldest son Adrian for his assistance with theoretical academic problems — I can only repeat 'Thank you!'

*James G. Rizzo*
*Lija, MALTA, July 1985*

# INTRODUCTION

This book is about model Stirling engines for beginners, with the emphasis on 'model' and 'beginners' to this fascinating hobby. There are not all that many books on the Stirling engine, in fact just enough to be counted on the five fingers of one hand, although there is an enormous wealth of contributions, technical articles, papers etc spread over several countries in different languages. A model engineer, a starter, who wants to learn more about the knack of building and running a model engine has to spend many hours researching before getting down to the workshop stage — some books just touch on the subject while articles with detailed projects have to be looked up and sifted for the right type of beginner's model.

This book covers the historical background of the Stirling engine, and gives a brief and elementary approach to the principle and some of the more important parameters of the hot air cycle. The second part of the book deals with projects that can be constructed easily (and cheaply) in a modest-sized home workshop, the projects increasing in sophistication and technical know-how. The emphasis here is on 'cheaply', because this needs not be an expensive hobby. Over 90 per cent of the engines built to date by the author have been constructed mainly from materials that came originally from scrap or surplus supplies. Nor is any great engineering skill required.

The thermodynamic principles of the cycles have been treated superficially for two reasons: the theoretical approach, with its explanations, is more complex than this book has been intended, and secondly there is little use for the beginner to be saddled with theory when the aim of the book is to introduce the Stirling engine in model form. From the author's experience this is a subject, which like other forms of engineering, is best treated in stages. One may experiment, and, having successfully completed one or more projects, decide to delve deeper into the subject.

Most of the modellers known to the author have started on this subject simply to satisfy a curiosity about whether such a contraption, without steam, petrol or electricity can really work. Most went on to build more sophisticated models, but some went on to experiment with certain principles or parameters to the benefit of greater knowledge to themselves and others. A few have successfully developed larger engines and quite a number of new ideas and mechanisms have been patented. Their interest invariably started from the small model hot air engine. One thing that the author has found in common — their enthusiasm to share their knowledge with other modellers and to help 'starters' iron out difficulties and solve problems.

CHAPTER 1

# WHAT IS A STIRLING-CYCLE ENGINE?

Basically a Stirling-cycle engine is a heat engine, that is an engine that derives its power directly from heat. Another name by which such an engine is known, the hot air engine, probably better describes the main characteristic of this type of engine — an engine which works on, or by means of, hot air.

## HISTORICAL BACKGROUND

The Stirling-cycle engine bears the name of its developer, the Reverend Robert Stirling, a minister of the Church of Scotland, who with his brother James, an engineer by profession, developed the hot air engine and endowed it with an invention which in those days was very far advanced in concept.

The Rev Stirling registered his 'invention' in 1816 (Patent No 4081) but failed to follow up the registration and to enrol the specification, so his application lapsed. His first 'hot air' engine, constructed in 1818, was quite successful and it is recorded that this engine, used as a quarry pump, ran continuously for over two years until it developed metal fatigue due to poor heat-resisting metals available at the time.

The earliest hot air or 'caloric' engine on record was introduced by Sir George Cayley in 1807 and patented in 1837. This design was improved by another engineer, a Mr Buckett, the redesigned engines being manufactured by the Caloric Engine Company. This type of engine (fig 1.1) as well as others

Fig 1.1 *The Buckett caloric engine* (The Engineer).

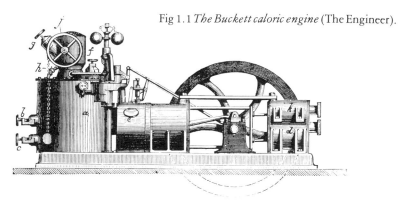

similar in design cannot be described as a pure hot air engine however, since a second and foreign matter, such as steam, gunpowder, gassified tar or turpentine oil, was employed in the explosion process.

The Rev Stirling's novel ideas were advanced in both concept and brilliance. His 1816 specification, *Improvements for diminishing the consumption of fuel,* makes excellent reading. His idea of regeneration, whereby heat was retained and re-used by the construction of 'tubes, passages and plates' made of 'metal, and any other substance that conducts and transmits heat easily', was an unusual one which was not accepted for several generations. Its real purpose could not be understood, simply because it was so much in advance of scientific knowledge of that era. In retrospect, it can be considered as one of the most amazing inventions ever.

For almost a century the Rev Stirling and his brother James were credited, erroneously, with two patents, one in 1827 and one in 1840. The 1816 specifications remained undiscovered until 1917, when the original documents were found and handed in to the Stirling family who passed them on to the Patent Office some 100 years after the date when they should have been enrolled in the official *Blue Book* series.

The Stirling engine of 1818 (fig 1.2) had a displacer (known in those days as 'transferer') and a power piston in the same cylinder. The one patented in 1827 (fig 1.3) was a twin-displacer engine where the power piston was double acting, at the upward stroke receiving air in the lower half from the displacer cylinder on the left, and air in the upper half from the displacer cylinder on the right on the downward stroke. A fourth small cylinder, connected by a system of levers, was used as a pressure pump. This engine was modified several times in later years to provide better regeneration, better cooling and less dead space.

**Below left** Fig 1.2 *Stirling's first engine, 1818* (The Engineer).
**Below right** Fig 1.3 *Stirling engine of 1827.*

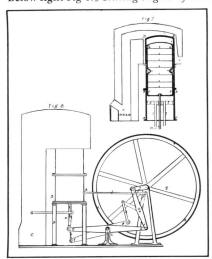
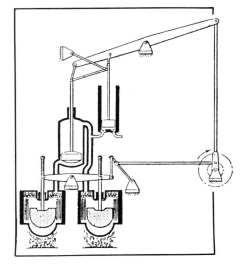

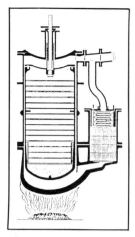 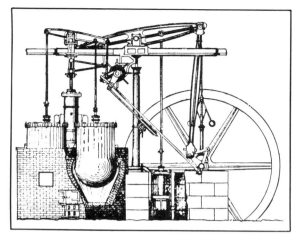

**Above left** Fig 1.4 *Detail from Stirling specification of the 1840 engine.*
**Above right** Fig 1.5 *Converted Stirling engine in a Dundee factory, 1843.*

The engine patented in 1840 was an improvement on the 1827 model particularly where the regenerator was concerned. In this engine the regenerator was removed from around the displacer cylinder and placed in a separate container with a cooler on top of it. Heated air from the bottom of the displacer cylinder passed through the regenerator, through the cooler and on to the power cylinder. The engine in fig 1.4 shows in a compact form the main principles of Stirling's invention and illustrates better than any other type of engine the construction of a near-perfect heat engine. It shows the source of heat (the furnace), the source of cooling (the water pipes) and between the two, the regenerator which abstracts heat as the air passes on its way to the cooler and returns heat to the air on its way to the source of heat.

Fig 1.5 illustrates a converted steam engine in a Dundee foundry which started to run in March 1843. It appears to have been a satisfactory experiment for about two and a half years, but in December 1845 a cylinder bottom overheated and failed. It was repaired twice and twice burnt out again. In the end the owners reconverted the motive power back to steam.

Hot air engines, known also as external combustion engines, can be broadly classified into two classes — closed-cycle engines where the working gas is used over and over again and open-cycle engines where a fresh intake of the working gas is admitted in the cylinder during each cycle and then discharged as exhaust. The Stirling engines are of the first type, that is closed-cycle engines. Most of the development that took place in later years was on this type of engine, although there were some exceptions, notably the Ericsson engines. However, few of the engineers who designed their engines on the closed-cycle included the regenerator. This was mainly due to the fact that they could not properly understand the function of this part! Open-cycle engines, on the other hand, required a system of valves with levers, links or cams to move them. Invariably these valves opened directly to the hot

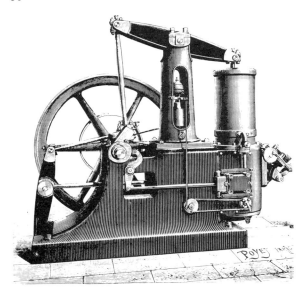

**Left** Fig 1.6 *Bernier's hot air engine—very popular in France* (The Engineer).

**Right** Fig 1.7 *An early Bailey pumping hot air engine* (The Engineer).

**Far right** Fig 1.8 *An early Robinson hot air engine* (The Engineer).

chamber and just as invariably these valves got burnt out with regular monotony. Obviously these engines were also very noisy.

Although the Stirling brothers patented three types of engines, it appears that they preferred the later separate twin cylinder for the power piston and displacer to the first type where the displacer and power piston work in one common cylinder. Yet subsequent development showed that this second type of engine is in fact the more efficient type. Other types of engines suffered from a major defect by using separate cylinders, in that part of the enclosed air, known as the dead space, performs no useful work and so reduces the specific power of the engine. For a while it was thought that the hot air or caloric engines would replace the steam engine which was going through a difficult time and still considered a dangerous piece of machinery, claiming countless lives as a result of boiler explosions.

Quite a number of caloric engines were developed and some firms established themselves well in this market for several decades. Two household names of the 1860s were the Bailey pumping engines and the Robinson domestic motors both of which were manufactured in great numbers. Several different sizes of Bailey engines were produced from ¼ hp to 3 ½ hp, none of them really powerful because of primitive construction and lack of regeneration. The only thing in their favour was the fact that they did not need any skilled labour to look after them and could be relied upon to work non-stop for weeks on end with only the supply of heat.

The Robinson domestic motors were more popular, being small in size and a very simple source of power. The first models, built in the 1860s, were rather crude, without regeneration and low in power. Later models incorporated a regenerative displacer, the displacer hollow body being filled with a mass of thin wire, through which the hot air flowed on its way to the power piston, placed at right angles to the displacer cylinder. Two features

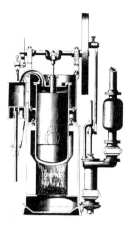
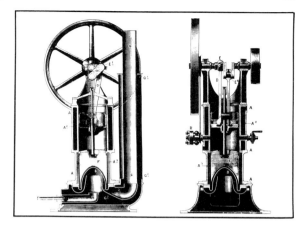

were outstanding in these later models; the short squat type of displacer and the displacer roller-drive off the power connecting rod which lessened the number of working parts. A model of this engine with a cut-away section of the displacer can be seen in The Science Museum, Kensington, along with a few other types of hot air engines, the most interesting being one of the early Philips air engines.

Between the 1860s and the 1880s a number of engines were manufactured in Europe and the States. One of the more successful early engines was produced in Europe by Lehmann. This design was not very different from the first engine built by Stirling, the notable difference being that the linkage mechanism was more compact, but no method of regeneration is apparent from sketches made at the time. It is probable that the unusually long displacer served also as a regenerator. The weight of this long displacer was supported by rollers. In the 1870s Stenberg of Denmark (Helsingfors) brought out the 'Calorisca' range based on Lehmann's design but with an

Fig 1.9 *Popular Robinson domestic motor* (The Engineer).

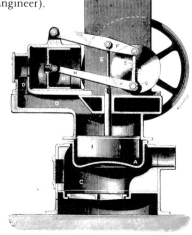

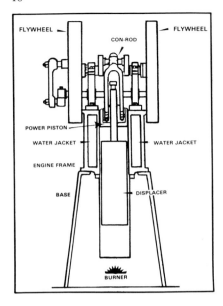

FLYWHEEL — FLYWHEEL
CON-ROD
POWER PISTON
WATER JACKET — WATER JACKET
ENGINE FRAME
BASE — DISPLACER
BURNER

**Left** Fig 1.10 *Artist's impression of a Heinrici engine.*

**Right** Fig 1.11 *Improved Rider hot air engine* (The Engineer).

**Bottom** Fig 1.13 *Ericsson's hot air engine* (The Engineer)

**Bottom left** Fig 1.12 *Rider's hot air engine* (The Engineer).

improved mechanism. At the turn of the century, and in the range of small domestic motors, as this size of hot air engines was called, the Heinrici motor (fig 1.10) appeared, achieving what was probably the greatest sale of hot air engines in Europe. This famous 'legs-in-the-air' engine had a simplified link drive which was quite graceful and attractive and also quite efficient for its size, (about 2½ft to 3ft tall). A large number of these motors were produced and exported to most European countries, and also to Asia. The construction was sturdy and the materials used were, for the time, of very high standard. Some of the engines in existence can still be made to run efficiently with a minimum application of heat. A similar version was brought out in England in 1908 by Gardner Ltd, but by then the era of the hot air engine was on the wane.

Turning to the United States, a name that became almost a household word was that of A.K. Rider of Philadelphia who manufactured and sold a brand of hot air engine, bearing his name, by the thousands. The Rider engine was an ingenious little hot air engine, compact and handy, ideal for domestic use particularly for pumping water. It includes almost all the features of the early Stirling engine except that where Stirling was careful to keep the power cylinder cool, even in the single cylinder engine, in the Rider engines the motive power is not only generated, but produced and applied in the hot cylinder. The two parallel cylinders are connected by a passage holding a regenerator consisting of a number of very thin iron plates packed closely together.

The Rider engines became very popular, not only in the United States where they sold by the tens of thousands, but also in England where they were manufactured, under licence, by Hayward Tyler & Co Ltd. The engines were used mainly to pump water, the pump being an important accessory to the

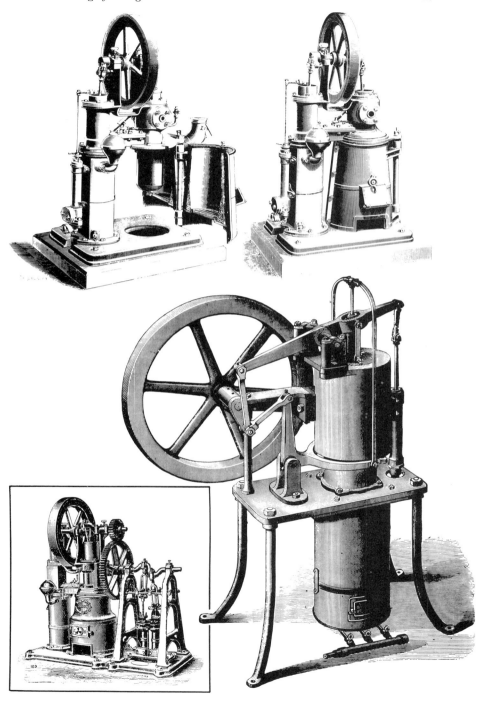

engine since the pumped water was directly utilised to cool one of the cylinders before it was discharged to serve its real purpose. Three sizes of Rider engines were manufactured, a 6 inch engine with a 125ft depth draw, an 8 inch engine with a 300ft depth draw and a 10 inch engine with a 500ft depth draw.

Ericsson reviewed his construction methods and brought out a single-cylinder hot air pumping engine of the closed-cycle type without any external valves, specifically designed as water pump. As in the Rider engine the pump was an integral part of the engine; the early models using the water flow to cool the upper part of the cylinder before discharging. Later models, of the 8 inch and the 10 inch size, had two pumps working from the main beam, a deep well pump and a surface pump, the former discharging into the water system while water from the surface pump was used first to cool the cylinder and then as hot water for domestic purposes. Four sizes of Ericsson hot air pumping engine were available, the above two sizes and two smaller ones, a 6 inch and a 5 inch domestic motor. Both the Rider and Ericsson engines were advertised as having interchangeable furnaces with the capability of using gas, paraffin, wood or coal depending on the client's choice.

There were several other names linked with the development and production of different types of hot air engines of both open and closed cycle type but few of these were of any real significance. The man who injected any stimulating development during this period was John Ericsson with his 'Sun Motor' (fig 1.14) which he built in 1872 and continued to develop until 1883 when it appears that interest was lost due to low efficiency and intermittent power output. This engine was a single cylinder concentric type with the displacer directly heated by a parabolic mirror concentrating the sun's rays on the hot end. A subsequent development appeared to have been planned in 1908 in Pheonix, Arizona, USA, where a Rider engine received solar energy from a concentrator, this being revolved to catch the sun's rays for most of the peak daylight hours. No official report of the project is known but it is presumed that it failed, most probably due to the small size of the concentrator.

During its relatively short period of useful life before it was superseded by the steam engine and later by the internal combustion engine, Stirling-cycle

Fig 1.14 *Artist's impression of Ericsson's sun motor.*

hot air engines were used in a variety of ways and for a number of industrial purposes. Apart from extensive use to pump water from mines, wells and rivers for which these engines were highly suitable, they were also used in workshops as prime movers. That is, driving ancillary machines by means of belts, in printing presses and in some known instances for dentists drills. Up to a few years ago table and stand fans, working on this principle, were very common and could be found in many homes in the United States.

The old drawings and woodcuts of these Stirling engines and the other hot air engines are worth studying. One cannot help but admire the ingenuity of the great engineers of the last century, who had ideas far in advance of their times, yet had to battle and work with inadequate materials and tools.

## SUBSEQUENT DEVELOPMENT

The story of the hot air engine from the late 1930s becomes not unlike a fairy tale with the charming prince in the shape of Philips of Eindhoven in the Netherlands bringing to life with a kiss, the sleeping beauty, the Stirling engine. This success story cannot be told briefly. What is important to this review is that while looking for a small, simple and light engine to generate electricity in remote areas for radio receivers and transmitters, Philips chose to investigate and research into the Stirling engine.

The first task undertaken by the Philips scientists was to analyse the defects in the Stirling engine in order to improve its efficiency to a standard comparable with the internal combustion engine. The engine design, the construction of the regenerator, the problem of heat transfer, resistance of the air-flow and the mechanical drives were all closely studied and the results incorporated in the first of a series of single cylinder engines combining all the major improvements.

The single cylinder engine has in the lower part of the cylinder a power piston through which a rod moves the transfer piston (displacer) in the upper part. The hot space, the regenerator, and the cold space are annular in shape and built around the cylinder. This configuration allows the engine to be very compact. The engine has a closed crank-case containing a small pressure pump connected to the crank and used to elevate the internal working pressure of the gas. The progression from this single-cylinder to the multi-cylinder engine would have been a natural one in the normal course of development. However in the case of the multi-cylinder, particularly in four-cylinder engines, the progress was simply amazing. The Stirling engine, with its double-acting capacity, lent itself much more to this type of configuration than to any other form (see Chapter 2). In 1970 Philips developed a four-cylinder in-line engine with their famous rhombic drive (see Chapter 6), developing about 100 bhp at 3,000 rpm. One of these experimental engines was fitted to a DAF coach and used with normal conventional transmission to drive the vehicle on the road.

Philips involvement and success aroused interest in a number of other firms. Moreover, Philips' attitude to share their knowledge made it possible for engineers and model engineers to take another look at the hot air engine. In fact the air motor became a possibility for anyone who had modest

Fig 1.15 *Philips engine under test* (Philips).

technical ability and construction facilities. Among the interested larger combines, General Motors, USA, developed in conjunction with Philips, a 3KW generator. They based on this engine a hybrid electric car powered by batteries and charged by the Stirling electric generator. Another large combine, the MAN-MWM Stirling Engine Development Group, jointly established by two major German diesel engine developers, joined Philips in building other experimental Stirling engines. United Stirling of Malmo, Sweden, was formed and licensed with the sole aim of manufacturing Stirling engines. This company, which has gone into development in a big way, is now producing large 200 hp double acting engines. In 1972 the Ford Motor Company started a joint development programme with Philips to develop a passenger car engine of about 175 hp. This was planned as a four-cylinder double-acting swashplate engine. At the same time Ford and United Stirling investigated another type of hot air engine of a slightly different technique. By 1976 Fords had installed a Philips engine in a Ford Pinto and a United Stirling engine in a Ford Taunus.

Of interest to those who believe in the future of the Stirling engine is a report, prepared by a team of engineers and scientists funded by the Ford Motor Company, to investigate the technical feasibility of heavy investment in the production of an alternative power unit for the motor-car. The report concludes that the Stirling engine, at a development cost of $8000 million would save a minimum amount of two million barrels of oil per day by 1999. Of even greater interest and even more technologically feasible in the field of automotive propulsion are the hybrid-electric car units. This system will certainly gain more interest as car batteries increase in efficiency and decrease in size and weight, making it possible for a medium sized hot air engine to give the necessary topping up charge to a family size round-about-town electric car.

## LATEST DEVELOPMENTS

Directly related to the Stirling-cycle engine have been at least four developments of interest to the model engineer. These are well within the means of an engineering society, a technical institute workshop or a higher education laboratory.

**The Beale free-piston engine** is a single cylinder engine without a crank case, crank-shaft or flywheel. This remarkable invention has only two moving parts, eliminating much hardware and hard work. It is simple in operation, compact, extremely quiet, and self-starting. This interesting development is the invention of Professor William Beale, of Athens, Ohio, USA. He made this discovery while professor of mechanical engineering at the University of Ohio.

The engine requires heating and cooling. Between these two actions exists the dynamic equilibrium of three forces exerted within a sealed unit, pressure forces, spring forces and damping forces. The operation involves a heavy and accurate piston and a displacer. The piston, in combination with an effective gas spring, oscillates at a natural frequency when heat is applied. This type of engine requires absolute accuracy in construction.

Beale engines are being developed for converting solar energy into electricity, total energy systems and water pumps amongst other developments. Although sealed, work can be extracted from the engine by attaching a load such as a linear electric generator to the reciprocating piston.

**Solar engines.** Stirling engines driven directly by solar energy must, of necessity, be of great interest to countries where energy demand is high and solar energy is a ready commodity in unlimited quantity. The fact that for two-thirds of the day such engines are inoperative should not deter their development but rather serve as an impetus to find ways of converting the energy either into stored energy such as batteries or stored potential energy for example the use of water pumped to higher levels and used to power hydraulic generators on demand.

In terms of large scale development there has been no substantial breakthrough. The Beale engine mentioned above is an excellent example of the type of converter required to service this shortcoming. Whether this development can be extended scale-wise to be of any interest to developing countries is another matter.

In the field of model engineering, on the other hand, the scope is unlimited, once an enthusiast has become proficient in the construction of small engines, particularly single-cylinder co-axial engines. With the aid of parabolic mirrors or Fresnel lenses sufficient heat can be generated at the hot end to enable the engine to develop relatively high speeds. Solar Engines of

Pheonix, Arizona, has produced commercially a miniature sun motor (some 8in long) powered by a parabolic mirror, of 18in diameter, which when aimed at the sun produces a speed in excess of 1,000 rpm.

**The thermo-mechanical generator (TMG)** is the first of two other important developments of the Stirling-cycle that should be described and considered here. Both are unconventional in the sense that they are a departure from the flywheel/crankshaft mechanical drive arrangement, yet in another sense both emerge from the same principle. Both developments originated from the Atomic Energy Research Establishment, Harwell, England.

The TMG was designed to generate low wattage electricity for equipment in remote locations where frequent maintenance and repairs are difficult. Therefore the prime requirement of this device is a continuous, efficient and highly reliable generator supplying electricity from a few tens of watts to a few hundred watts, depending on the equipment to be powered, whether it is a sea-buoy or a weather station radio, for long periods without servicing.

The main secret of the success of the thermo-mechanical generator lies in the restriction, to a large degree, of friction losses. A 'normal' hot air engine works by volumetric changes in gas as a result of temperature changes brought about by displacing a fixed amount of gas to and from hot and cold chambers; to perform this work such an engine uses a displacer and a power piston to shift the gas from one end to the other. The thermo-mechanical generator uses a displacer and a semi-fixed substitute power piston to do the equivalent work. A linear alternator is necessary for the generation of electricity since the movement of the diaphragm is too small to be coupled to the crankshaft of a rotating engine.

A 120 watt version of the thermo-mechanical generator, is currently under development by Homach Systems Ltd of Swindon, with a re-designed diaphragm system. This uses an articulated diaphragm invented by Mr. E.H. Cooke-Yarborough, one of Harwell's leading scientists, now collaborating with the above firm. The TMG has created a considerable interest in technical and scientific circles in view of its compact design and trouble-free maintenance.

**The 'Fluidyne' engine pump** (fig 1.16) is the other development that came originally from the Atomic Energy Research Laboratory and it too is related to the Stirling-cycle family. In the Fluidyne there are no metal working parts as we know them. No displacer, power piston, crankshaft or flywheel. Yet this 'motor' can be made to pump water in a steady rythmic flow when heat is applied to one end and the other end is kept cold. In principle it is a remarkably simple device and relatively easy to build in model form, in a laboratory or workshop.

The basic engine consists of two tubes containing a column of water topped by a pocket of air. The columns are joined at the top by an inverted open U-tube containing air (as the working gas) and joined at the bottom by a long tube connected to the pump body, a vertical tube which acts as the pump and encloses the only mechanical items, an inlet and an outlet ball valve. The pump sucks water through the bottom valve and expels it in spurts through the top valve as the gas in the interconnected air space above the hot and cold chambers alternately expands and contracts. The Fluidyne engine pump runs

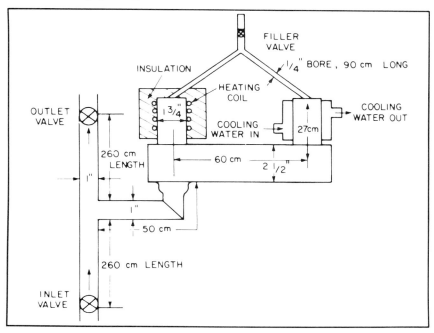

Fig 1.16 *Fluidyne engine* (Dr Colin West).

as long as heat is applied. The Fluidyne has a low thermal efficiency, but, conversely, since it can be powered by low cost heat, such as waste pipe heat and readily available crude combustible material, the efficiency factor does not matter much. The Fluidyne pump can be used in third-world countries powered directly by free solar energy and that is where its real potential lies.

While the Fluidyne was being developed at Harwell, good results were obtained with this ingenious device, capable of pumping about 400 gallons per hour to a height of 10 ft with an efficiency of 4.7 per cent. Later, under development by the Metal Box Co of Calcutta, India, it was reported to be able to pump 2500 gallons of water per hour at a head of 10 ft with an efficiency of 7 per cent.

The Fluidyne was devised at the AERE, Harwell by Dr Colin West, who has written a very interesting book on this subject, *Liquid-piston Stirling engines*, published by Van Nostrand Reinhold Co of New York, USA.

## FUTURE DEVELOPMENT

There now appears to be a good future for the Stirling engine. The use of heat resisting alloys and ceramics, better sealing techniques and the use of computers are all contributing factors in making this engine more efficient. Development in the recent past has been rather slow due to the fact that for many years it was more expedient to subsidize research into petrol and diesel engines. Now that these engines have been fully developed it appears that it

Fig 1.17 *A multi-fuel demonstration of the Philips hot air engine, showing the capability of the engine to work equally well with different types of fuel. The fuels used on the test bed engine are: alcohol, olive oil, salad oil, lubricating oil, fuel oil (two types), kerosene, diesel oil, petrol.*

may be the turn of the Stirling engine to be given its proper place as a prime mover.

In many ways the Stirling engine is superior to other engines (steam, petrol and diesel). These reasons may be summarised as follows:- **1** The Stirling is relatively less noisy. In many cases it is considered to be a quiet engine. It is also comparatively free from vibrations. **2** It creates no pollution problems. Since heat is applied externally there are no fumes or contamination and the exhaust or smoke is relatively clear. **3** It runs on a variety of fuels: oil, wood, coal, paraffin, alcohol, gas, straw, sawdust and compressed paper are a few of the fuels which have been used. Also, it is possible to change from one fuel to another without practical difficulties. **4** It is more economical to operate. Some of the fuels that can be used to generate heat are cheaper to obtain than oil-based products. The Stirling engine needs very little lubrication and this is also a substantial saver over a period of time. **5** There are fewer moving parts in Stirling engines than in petrol and diesel engines. Moreover, there is less stress on the moving parts. These two factors combine to give longer life to these types of engines.

CHAPTER 2

# HOW THE HOT AIR ENGINE WORKS

When a pocket of air or gas enclosed in a cylinder is heated, the pressure inside the cylinder rises; if one end of the cylinder can slide, the increased pressure will push that end until the increased volume is accommodated (fig 2.1). If the same volume of hot gas is cooled suddenly, the gas contracts, the pressure falls and a vacuum is created. The sliding end, if tight fitting, will be sucked in (fig 2.2). If the sliding end is made into a tight fitting piston attached to a crankshaft with a flywheel, and the device is again subjected to the above changes of temperature, the volumetric changes will produce motion of the flywheel (fig 2.3). That, basically, is the principle of the hot air engine. Since it is not physically possible to heat and cool gas rapidly and in quick succession, another device is employed which in practice does the same work. This device is called a 'displacer'; In Stirling's days it was called a 'transferer', which probably describes its actions better. The work of the displacer is to shift the volume of gas from one end of the cylinder to the other. Again, if the cylinder is heated at the closed end and cooled at the other end where the piston is, and if the displacer is, by mechanical means,

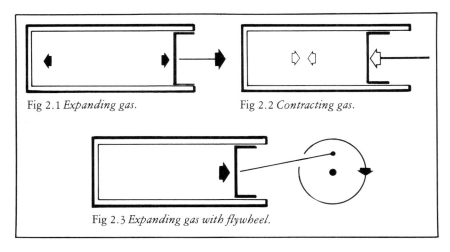

Fig 2.1 *Expanding gas.*    Fig 2.2 *Contracting gas.*

Fig 2.3 *Expanding gas with flywheel.*

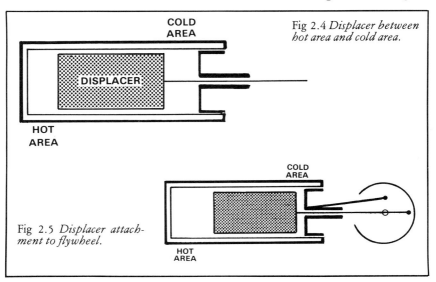

Fig 2.4 *Displacer between hot area and cold area.*

Fig 2.5 *Displacer attachment to flywheel.*

made to shift the gas from one end of the cylinder to the other, the gas will alternately expand and contract depending at which end it is (fig 2.4).

Thus, if the displacer shifts the gas to the hot end, gas will expand. When the gas is shifted to the cold end, it will contract. Every time the gas expands and contracts a cycle takes place. As a result, the piston at the open end will slide out and then return to place. Obviously flywheel momentum is necessary to push back the piston after completing half a turn, as well as to make the movement smooth (fig 2.5). The displacer is loose fitting, its work being solely that of shifting the gas. If the fitting is too tight, gas will not be able to move easily from one end of the cylinder to the other. If it is too loose fitting, gas will escape along the annular gap; therefore the fitting of a displacer is somewhat critical.

In simplified terms the whole cycle takes place in this manner: **1.** Gas is heated at the hot end of the cylinder; the increased pressure forces out the power piston as far as it can go. **2.** Gas is shifted by the displacer to the cold end where it contracts and decreases in volume; the power piston is pushed back by the flywheel mechanism and, helped by the vacuum created, 'sucks' the piston. **3.** Gas is shifted back to the hot end, heated, expands and the cycle is repeated.

The displacer and the power piston do not move backwards and forwards at the same time; rather there is a calculated gap (both in time and in motion). This gap works out at a quarter of a turn of a circle or as it is more commonly known, a 90° phase; the stroke of the displacer always leading by approximately 90° (or a quarter of a turn) ahead of the power piston. The work of the displacer is to transfer gas from one end of the cylinder to the other. The work of the power piston is to utilise the compression of the gas at low temperature and the expansion of gas at high temperature, and to transfer this movement by means of a crankshaft to produce motion.

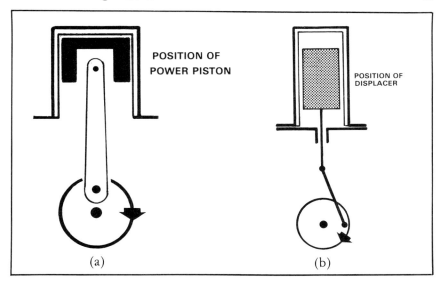

Fig 2.6 *Positions of power piston and displacer (90°) (a) The power piston has reached the top of its stroke, known as TDC (Top Dead Centre), and has exhausted its power phase. (b) The displacer is a quarter of a turn ahead, already shifting gas from the cold end to the hot end in preparation for the next expansion phase.*

While the work of the engine appears to be a result of gas being pushed backwards and forwards, at the same time being heated and cooled, it is not easy to comprehend what is actually happening inside the engine, and no amount of theory or diagrams can really give a visual insight to the cycle. Unlike the internal combustion engine where each cycle of operation consists of four distinct and clearly defined phases, namely: induction, compression, combustion (ignition or power) and exhaust; the hot air engine's phases are not so distinct and clear cut, with one phase leading to another. A continuous movement of the working gas is taking place between the closed side (hot end) of the cylinder and the side closed by the power piston (cold end). There is no sharp transition between the successive phases. The clearest picture one can get of the whole cycle would be to follow these four stages, on the understanding that there is no first and last stage and that one stage follows into the next in a continuous motion without any transition (fig 2.7).

Two basic factors determine the efficiency of the Stirling-cycle engine. These are the sealing of the gas inside the cylinder between the closed end and the power piston end and secondly, the temperature difference between the hot and cold spaces of the cylinder.

**Sealing** affects efficiency in the following way. When gas inside the cylinder expands, the increased pressure pushes back against the side which gives the least resistance — in this case the power piston. The power piston in turn gives way as far as possible without allowing the gas to escape out of the cylinder. Technically and ideally, the amount of gas inside the cylinder should not alter. In practice, however, some gas does escape. When that happens the

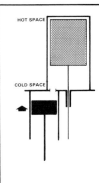

**Left** *In stage 1, the displacer has pushed the working gas into the cold end of the cylinder, where it is cooled. The pressure at this stage is normal. The power piston is being pushed inwards by flywheel momentum helped by a partial vacuum created by the cooling gas.*

**Right** *In stage 2, the displacer is pushing the gas into the hot space and already a certain amount is being subjected to heat. The power piston is at TDC, awaiting increase in pressure as a result of expanding gas.*

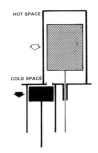

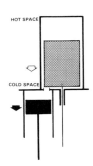

**Left** *In stage 3, the displacer has pushed all the gas into the hot end, with a corresponding increase of pressure to the maximum. The power piston is being pushed by the increased pressure and is applying force to the flywheel, thus creating work.*

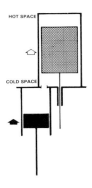

**Right** *In stage 4, the displacer is pushing the gas into the cold space where the pressure will fall and a partial vacuum created. The power piston has reached its maximum stroke and is ready to travel back to TDC under flywheel momentum and the sucking action of the falling pressure (partial vacuum).*

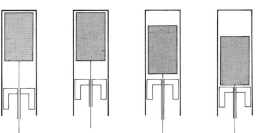

**Left** *The single cylinder co-axial engine has an almost identical movement of the displacer and power piston with the one difference that they both sweep part of the cylinder (ie, common to both) although at different times.*

Fig 2.7 *Top: Stages 1 & 2 of working engine. Centre: Stages 3 & 4 of working engine. Bottom: Positions of co-axial engine.*

cycle efficiency drops and if the gas leak is substantial the engine fails to run, or if it runs, it does so for only a few minutes. It is therefore very important that the power piston should be gas tight.

**Temperature difference** also effects efficiency. The alternating expansion and contraction of gas (together with the increase and decrease of pressure) inside a cylinder is the real working force of the Stirling engine. This working force can only be effective if there is substantial difference in temperature between the hot end and the cold end of the cylinder. Therefore the higher the temperature at the hot end and the colder the temperature at the cold end, the more efficient the working force becomes. Both heating and cooling however create problems if they are to be carried out to extreme temperatures and for long periods. For example, most metals in common use, including steel, will distort if heat is applied to a high degree and for a long time. Research has been going on to find an alloy which can resist extreme temperatures. Ceramic has also been mentioned as a possibility. An engineer who is experimenting on model engines however, may use steel without much difficulty and with good results.

It is relatively easier to provide cooling. Exposing a heated area to the surrounding atmosphere is one way. This is, of course, a slow process which can be hastened by enlarging the surface transfer area. The use of a cooler, whether a solid such as fins, or a part solid/part liquid such as a cooling tank, accelerates the dissipation of heat. Quite apart from this, the surrounding metal, cylinder, pistons, con-rod and regenerator (if any) together absorb an appreciable proportion of the heat. Methods of providing heating and cooling systems are discussed in Chapter 4, but there are also many other factors which have a direct bearing on the efficiency of the model or small engine.

**Regeneration** of energy plays a great part in the design and construction of an engine. Heating and cooling the working gas several hundred times a minute is no mean feat for a small engine particularly as the extremes of temperature can be a couple of hundred degrees or more apart. This work can be made easier by an arrangement whereby the hot gas on its way to the cold end loses its heat gradually while the cold gas, on its way to the hot end picks up heat along the way in order to reach the desired temperature and the necessary expansion in the hot chamber in the minimum of time, thus allowing for greater speed in the transfer of gas.

Two basic methods have been used in model engines to assist in the way hot gas gradually loses its heat and cold gas gradually absorbs it. The first uses the displacer to do this heat transfer. The displacer body is made sufficiently long for the hot gas moving along it to transfer heat to the body of the displacer in what is termed a 'temperature gradient'. The body becomes hot at one end and tapering off in heat progressively towards the other end. The second method is more ingenious and in some model engines very effective. This involves the building of a regenerator outside or around the main cylinder through which the gas passes. The same principle of heat absorbtion applies to this kind of regenerator. Heat is absorbed in a gradient manner by a mass of metal, which makes up the regenerator and this heat is picked up again by cold gas on its way to the expansion chamber. (Further details on regenerators in Chapter 3.)

Fig 2.8 *Philips rhombic drive mechanism* (Philips).

**Drive mechanisms** are a factor that can make or mar an engine. Many mechanical arrangements have been 'invented' to provide the necessary 90° phase between the displacers and the power pistons. The most common arrangements in old days involved complicated systems of levers and cranks in order to achieve this 'discrepancy' between the movements of the two pistons. The situation was complicated when the need arose to have different length of strokes for the two pistons.

Although the old engines are a delight to watch with all sorts of levers, rocking levers, cranks and bell-cranks going in all directions, the engines are slow, rarely exceeding 200 rpm and very often in the region of 100 rpm. Back in the 1850s to 1900s these engines were used mainly to pump water. High revolutions were not important for this task and speed could be sacrificed for reliability. Leverage tends to create friction and therefore designers of modern Stirling engines avoid complicated systems of levers and cranks.

One of the more brilliant and successful drives, the rhombic drive mechanism, was devised and perfected by Philips. With this method the displacer rod passes through the power piston rod without exerting lateral thrust on the power piston. It is a compact mechanism capable of sustained pressure. Another excellent design is the swashplate mechanism which drives a number of cylinders successively by a continuous circular motion. Such engines are designed to work horizontally and with a multiple of cylinders (usually four to six in number). There is still room for more inventiveness, however, when it comes to the mechanical arrangements of the Stirling engine and many a young engineer (and some not so young, too) have come forward with workable and commendable ideas. New drive arrangements are still being devised and some may yet become commercially viable. (Drive mechanisms are discussed at length in Chapter 6.)

## DESIGN VARIATIONS

Stirling engines can be broadly divided into two distinct groups; the single-acting engines and those that are double-acting.

**Single-acting engines** are those which enclose in each cylinder a piston and a displacer, or two pistons with an expansion space and a compression space. The most common basic arrangements for single-acting engines are shown in the accompanying diagram (fig 2.9). The V-type engine (fig 2.10) is popular with model engineers since the configuration makes the drive mechanism fairly simple to construct. Such engines have an outside regenerator.

**Double-acting** engines are those engines with two or more cylinders, the

Fig 2.9 *Configuration of engines. A and A1 — single cylinder co-axial configuration with the displacer and power piston working in the same enclosed space; also known as the Beta configuration (Prova II, Dyna and Sunspot described in this book, are built on this configuration). B and B1 — twin-cylinder configuration with the displacer and power piston occupying and working in separate areas — also known as the Gamma configuration. (Dolly, Lolly and Sturdy are built with this configuration.) C and C1 — twin-cylinder with power pistons each working in a separate space — also known as the Alpha configuration, known sometimes as the Rider engine after its developer. Some engines of this type are modified by the addition of an extension to the power piston on the hot side to keep the power piston at some distance from the hot cap.*

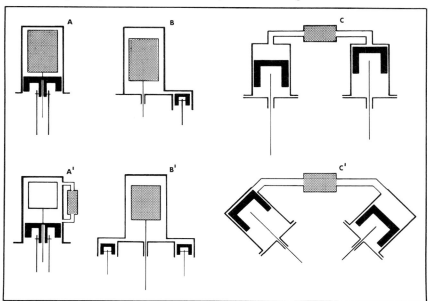

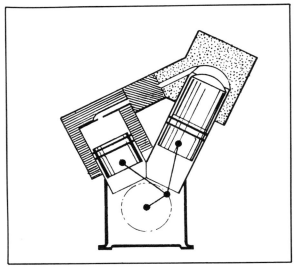

**Left** Fig 2.10 *V-type engine configuration.*

**Below** Fig 2.11 *Double acting configuration.*

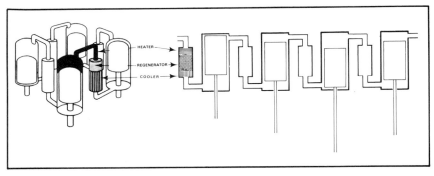

cylinders being connected in such a way that the expansion (hot) end of one is connected to the compression (cold) end of another, the connection passing through a regenerator. Each cylinder holds a single piston which serves as a power piston and displacer at the same time (fig 2.11). The main advantage of this type of operation is that the number of elements per unit is halved. Thus if in a single-acting engine with four cylinders there are four power pistons and four displacers plus complicated con-rod mechanism, a double-acting engine with four cylinders needs only four piston-displacers and a less complicated con-rod arrangement. The main disadvantage of the double-acting configuration lies in the lack of flexibility in design and the fact that one cannot experiment as with a single cylinder engine.

Manufacturers of the larger type of hot air engines are attracted by the advantages of minimum working parts, and current development is now geared to double-acting engines. In the field of model engineering and small engine production the single-acting engine will continue to attract the greater development in view of the simplicity of design and the greater tolerance that these engines have to changes in parameters and mechanical conditions.

## VARIATIONS OF THE STIRLING-CYCLE

**The thermo-mechanical generator** works on the general principle of a hot air engine with the following mechanical alternations. The displacer is retained but the power piston is replaced by a fixed, flexible metal diaphragm. The displacer is mounted on a flat circular spring attached to the main body casing which allows the displacer to move some $3/32$ in without touching the body or the power diaphragm. A propane gas burner heats the cylinder base to some 800°F and the engine self-starts automatically on the application of heat. The expansion of the gas in the main cylinder causes the diaphragm to expand and this in turn oscillates. The oscillation, together with machine vibration, causes the displacer to move vertically, thus displacing the working gas (helium) from the hot end to the cold end of the engine. The cold end is kept at a low temperature by a closed cycle cooling system (fig 2.12).

The flexible diaphragm, acting as a power piston, reciprocates the armature of a linear alternator and as a result of the electrical loading by the alternator, lags 90° out of phase with displacer. Since there are no sliding surfaces, the generator is efficient at power levels of a few tens of watts and because there is no static friction it is self-starting. Moreover there is no mechanical wear and no need for lubrication. All flexing components are stressed to well below the fatigue limit, so there is no known limit to the operating life. Indeed a test diaphragm ran continuously for two and a half years. When compared with alternative sources of storaged energy the advantages of such a generator can be seen. For example, it would need a 1½ tons weight of batteries to give a 25 watt power output for one year, a job for which the TMG is admirably suited.

**The Fluidyne** engine works on the principle of the Stirling-cycle. Heat is applied to one end of a cylinder, while the other end is kept cool; the working gas expands and work is performed. In this case, however, there are no metal parts, no piston, displacer, con-rod, crank, flywheel, etc. The working

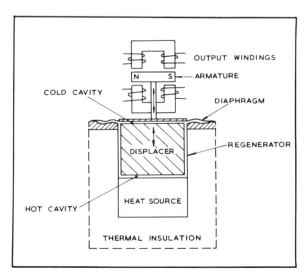

Fig 2.12 *AERE R-8036* TMG (AERE).

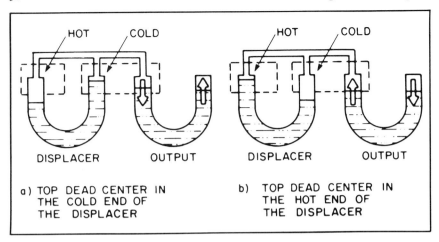

a) TOP DEAD CENTER IN          b) TOP DEAD CENTER IN
   THE COLD END OF                THE HOT END OF
   THE DISPLACER                  THE DISPLACER

elements are gas (air) and a liquid (water). The operation of the basic Fluidyne is described by Dr Colin West in his book with the aid of the accompanying illustration (fig 2.13) and the following description: 'Suppose the water in the displacer is set to oscillate from one limb of the U-Tube into the other limb and back; TDC in the cold end will correspond to the BDC in the hot end, as shown in the left hand part of the illustration, in which most of the air trapped above the water in the displacer is in the hot left-hand limb. Most of the air is therefore hot, so its pressure rises, tending to force the water in the output tube to move from right to left, as the arrow indicates. Half a cycle later the displacer water will swing back into the other limb, so that the cold surface is at BDC; this is shown in the right hand part of the illustration. Most of the air is now in the cold side of the machine, so its pressure will fall, pulling the water in the output column back from left to right!

If the output tube is attached to a vertical tube with two one-way valves, such that water pushed out of the output tube flows upwards (without return) then on the swing of the displacer, water will return to the hot end and the vacuum created in the output tube will lift water from a tank to replace that pushed out of the output tube. Each cycle will, therefore, alternately push out water from the top and suck water from below. The Fluidyne possesses some other attractive features. These are: low capital costs in manufacture, even with small production volume; simplicity, with ease of construction; durability and reliability; no solid moving parts; no lubrication required; use of low grade heat; starting on thermal-effects alone (self-starting); quietness of operation; long service life and high work output to cost ratio.

## REASONS FOR THE INEFFICIENCY OF EARLIER HOT AIR ENGINES

1 Early hot air engines were very clumsy; they had to be equipped with heavy moving parts which caused large friction losses and thus the mechanical efficiency was very low.

**Left** Fig 2.13 *Basic operation of the Fluidyne* (Dr Colin West).

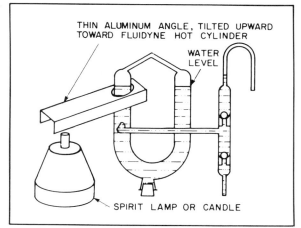

THIN ALUMINUM ANGLE, TILTED UPWARD TOWARD FLUIDYNE HOT CYLINDER

WATER LEVEL

SPIRIT LAMP OR CANDLE

**Right** Fig 2.14 *Small glass Fluidyne pump* (Dr Colin West).

**2** The metal walls of the hot cylinder, in order to be sufficiently strong to withstand constant high temperature, had to be made very thick; thus the heat conduction losses were tremendous.

**3** The temperature of the wall of the hot chamber was relatively low while the temperature inside the hot air chamber never exceeded 600°F.

**4** Both the heater (furnace) and the cooler (when used) were primitive and rarely was the large difference in temperature maintained for long periods. Later engines which had a continuous flow of water by means of pumps (such as the Rider), were more efficient than others.

**5** Many engines had a heavy transferer (displacer), often made from beaten iron sheets. The construction was such that after a relatively short period of heating, the transferer served as more of a heat conductor than as a regenerative displacer, thus lowering the temperature difference.

## SOME CHARACTERISTICS OF THE STIRLING-CYCLE ENGINE

**1 Starting.** Stirling engines are not self-starting; in fact it is one of the main disadvantages of these engines. Once the heated end of the cylinder(s) has reached the required temperature, force or push is applied to the flywheel which will then accelerate rapidly. In the case of small engines, a flick on the flywheel is sufficient — in the case of bigger engines a starter motor is required.

**2 Speed Control.** Certain Stirling engines are designed to maintain a constant speed whatever the load — these include electric generators and water pumps. Other engines may require speed variation. One method of increasing speed is by increasing heat, however the response of an engine to an increase in temperature is slow and this may only be suitable for some types of engines.

Engines used in the automobile industry are designed to respond rapidly to the power load; one of the methods used for faster acceleration is to increase the pressure of gas inside the cylinder. Another method is the use of valves to control pressure; bleeding pressure when slow speed is required and closing the valves and increasing pressure when fast speed is required.

Another method of varying the speed is by slightly varying the phase relationship which, in turn, alters the pressure. This method seems to have more scope for development, and it is the method on which experiments are currently taking place in several laboratories.

**3 Pressurisation potential.** A Stirling engine that operates at normal air pressure develops little power. If the engine is pressurised, however, the power increases greatly. An engine that works on helium or hydrogen and is, moreover, pressurised, will develop such power that is easily comparable to petrol/diesel engines when the ratios of weight to power are taken into consideration.

**4 Fuel versatility.** The Stirling engine has been called 'the future engine of the Third World'. When the supply of mineral fuels is very much an open question for the future and one considers that solar energy can provide an alternative 'fuel' for this type of engine, one can see the attraction that the development of the Stirling-cycle holds in the next decade.

CHAPTER 3

# THE REGENERATOR

The name originally given to what we now call the 'regenerator' was 'economiser' or 'heat economiser', which probably described better the function of the device. It is certain that the inventor, Robert Stirling, had in mind economy of energy by using the device to store up to 95 per cent of the heat given by the furnace and re-using it continuously. The wasted 5 per cent of heat was made up by the furnace. Therefore, using this method ensured a saving of energy—an 'economiser' of heat.

For many generations the principle of the regenerator was not understood by successive manufacturers of hot air engines — indeed most of the engines they built had no regenerator. Other manufacturers, like Ericsson, Robinson, Lehmann etc, experimented with and without regenerators. Eventually the more successful engines incorporated some version or other of the regenerator. The regenerator is used, with various modifications, in all hot air engines under development today. It is now accepted that a regenerator is the most essential component in a hot air engine and that the efficiency of an engine is largely determined by the regenerator. Yet little is known about what actually happens inside the regenerator and in spite of the many detailed studies, the best designs for efficient regenerators depend on the results of actual experiments.

## DESCRIPTION

The regenerator may be described as a 'sponge, alternately absorbing and releasing heat'. In its simplest form it can be a mass of metal between the two spaces of the hot air engine, the hot or expansion space and the cold or compression space. This 'sponge' of metal absorbs heat from the gas (which has been subjected to high temperature) on its way to the cold space where it is cooled, compressed and returned to the hot space (fig 3.1). On its way to the cooling agent, the hot gas leaves behind in the 'sponge', a considerable amount of heat in a temperature gradient, with a high temperature on the side of the hot space, cooling progressively towards the cold space (fig 3.2). In the compression stage the volume of gas which has been compressed, is cooled and then pushed back through the regenerator from where the stored heat is

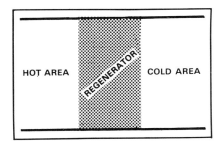

**Above** Fig 3.1 *Regenerator—a simple sketch.*

**Right** Fig 3.2 *Regenerator—temperature gradient zones.*

picked up (rather like the sponge being squeezed) prior to its entry into the heater.

In practice the regenerator works in the following manner. The heater brings up the temperature of the gas in the hot area to a temperature of, say, 400°F while the cooler maintains a temperature of 40°F in the cold space. The ideal regenerator will provide a temperature gradient of 300°F on the hot side and 100°F on the cold side. Gas entering the hot chamber from the regenerator is already at a temperature of around 300°F — therefore the heater has that much less work to heat the incoming gas from 300° to 400°F. Gas entering the cold chamber from the regenerator is at a temperature of around 100°F, having left much of its heat behind in the regenerator, and therefore to cool this gas to 40°F, the cooler has relatively less work to perform.

Without the regenerator, both the heating and cooling elements have far greater work to do, heating gas from 40°F to 400°, while the cooling agent cools the gas from 400°F to 40°F, all this as much as 1,000 times a minute if not more (fig 3.3). Therefore the regenerator, apart from appreciably lessening the work of the heating and cooling elements and thus saving energy, also increases many times over the efficiency of the Stirling-cycle. The four types of regenerator; external, external annular, internal annular, and displacer, are discussed below.

**External regenerators:** In this type of regenerator the gas leaves the main body of the cylinder, passes through an external regenerator back into the main cylinder or to an adjoining cylinder which is connected to the same drive mechanism. This regenerator was first used by the Stirling brothers in an engine built around 1840. In this engine, the regenerator (and the cooler) were housed in a separate cylinder connected to the displacer cylinder and leading to the power cylinder. In this cycle, the hot air was forced out of the displacer cylinder by the displacer (or transferer) through the regenerator and cooler into the power cylinder, compressed and returned via the regenerator into the displacer cylinder.

The regenerator used in this particular engine needs to be described in

Fig 3.3 *Regenerator—*
*temperature gradient.*

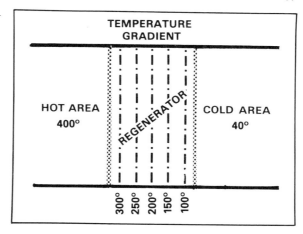

order to understand the ingenuity of the inventor. The regenerator consisted of four separate stacks, each identical in shape and construction. Each stack contained a quantity of iron sheets $1/40$ in thick, standing upright and separated from each other by ridges with passages in between of about $1/50$ in width. The stacks themselves were separated in order to reduce longtitudinal heat conduction. Dead space was reduced by the filling of passages and gaps with pieces of broken glass, which also served as part of the regenerator. Altogether, considering the lack of suitable materials in those days, and working more on intuition than expertise, this type of regenerator was very far advanced in concept (fig 3.4).

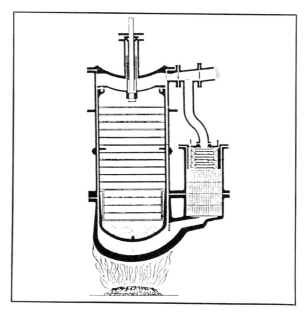

Fig. 3.4 *Stirling's regen-*
*erator of 1840.*

**External annular regenerators:** In this type of regenerator the working gas leaves the displacer cylinder, passes through a regenerative matrix built between the cylinder and an outside cap or cover and is channelled into the power cylinder through a cooler. The regenerator matrix is totally enclosed within the outer cap to prevent loss of compression or escape of gas, and is built in such a way as to reduce dead space to a minimum. This type of regenerator was first used by Stirling in 1827. A description of the engine is worth reading as it explains the principle of the regenerator or 'economiser' as was envisaged by the inventors (fig 3.5).

'Displacer **B** moves vertically within the inner cylinder, fitting easily and without friction. The bottom of the inner cylinder **X** is drilled with several holes. Air is pushed to the outer cylinder **YY** which is under constant high temperature, through the regenerator **C**, through the cooler **D**, through the passage **E** to the power piston **A**. The annular space between the outer and inner cylinders contains the regenerator **C** which is a grating composed of thin vertical oblong strips of metal with narrow passages between them. The cooler **D** consists of a horizontal coil of fine copper tube through which a current of cold water is forced'.

The operation was further described as follows: 'The effect of the alternate motion of the plunger (displacer) **B** is to transfer a certain mass of air, which may be called 'working air' alternately to the upper and lower end of the

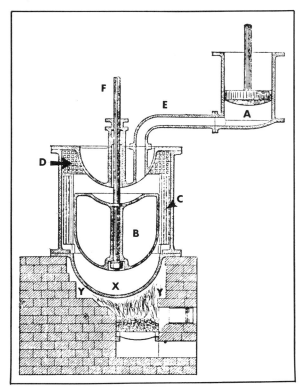

Fig 3.5 *Stirling's regenerator of 1827. A, Power piston, B Displacer, C Regenerator, D Cooler, E, Connecting air passage, F Displacer rod, X Inner displacer cylinder, YY Outer (and heated) cylinder.*

**Right** Fig 3.6 *Philips engine.*

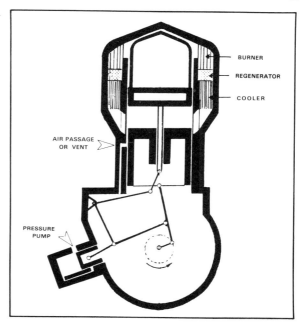

**Below** Fig 3.7 *Philips regenerator coils* (Philips).

(outer) cylinder, by making it pass up and down through the regenerator **C**. The perforated hemispherical inner cylinder **X** causes a diffusion and rapid circulation of the air as it passes into the outer cylinder and thus facilitates the convection of heat to it, for the purpose of enabling it to undergo the expansion which lifts (power) piston **A**. The descent of the plunger **B** causes the air to return through the regenerator to the upper end of the receiver. The greater part of the heat is stored in the plates of the regenerator; the remainder of that heat is abstracted by the refrigerator (cooler). The heat stored in the regenerator serves to raise the temperature of the air when this is returned to the lower end of the cylinder'.

The Philips air engine (fig 3.6) was constructed with the same style of regenerator between the displacer cylinder and the outer cylinder or cap. Philips, however, have used a much more efficient type of regenerator, consisting of a porous coil of thin metal wire (fig 3.7). It has been stated that this regenerator has over 95 per cent efficiency and is capable of raising the temperature of a quantity of air flowing through it from 100°C to 600°C within 1/1000th second and the reverse, that is cooling from 600°C to 100°C in the same fraction of time, while the temperature gradient in the direction of the air flow is maintained without appreciable loss of heat.

The construction of the award winning model engine 'Prova II' described in Chapter 13, involved the building, quite successfully as it turned out, of an external annular regenerator inserted between the regenerator cover and the engine cylinder. The first regenerator consisted of steel shim, 0.002 in thick, which was corrugated in shape and inserted to allow the working gas to pass on both sides. Further experiments with other regenerative material, such as fine steel mesh, slivers of ceramic and strands of fine steel wire, are taking place.

**Internal and annular regenerators:** This type of regenerator is contructed or fitted in a narrow annular gap between the displacer and the cylinder wall or between two cylinders as in the Rider engine (fig 3.8). The regenerator

Fig 3.8 *Rider engine and regenerator* (The Engineer).

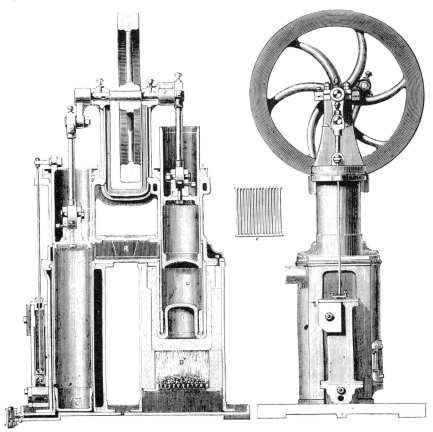

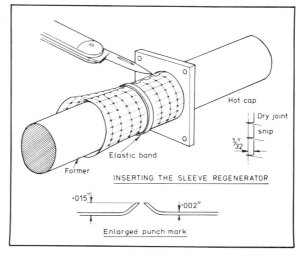

Fig 3.9 *Mr R. Robbins' 'Ergo II' sleeve regenerator made out of 0.0002 in. steel shim. This regenerator was used with great success in his award-winning 19cc V2 marine Rider engine.*

consists of thin steel sheet, steel mesh or coils of fine steel wire which allow heated gas to pass longitudinally, preferably on both sides, of the metal surface, depositing and withdrawing heat from the surface of the regenerative material. The steel sheet is usually hammered to raise dents or ridges on alternate sides, this method being used primarily to keep the sheet from surface contact with the cylinder wall and at the same time allowing gas to travel on both sides of the surface.

Occasional perforations permit airflow and create some turbulence, which is desirable. In the case of the steel mesh the only requirement is a minimal distance from the cylinder wall. This is by far the best type of regenerative material for internal regenerators since it has most of the important characteristics, low conduction, good airflow and turbulence and light weight.

Fine steel wire can be used for the purpose of regeneration as a coil inserted along the wall of the displacer cylinder or wound round a light-weight displacer. The first method was used successfully in the building of 'Dyna' (Chapter 15). The 1.5 mm steel spring wire was used to narrow the annular gap. Excellent regeneration and low conduction of heat enabled this engine to develop power and to run for long periods. The second method was used by Stirling in 1818 in his first engine which had an elongated displacer with several layers of thin gauge wire around it. The author has not heard of any model engine with such a regenerator, but such a construction could be successful on a model. The only problem envisaged here is that of weight. Engines using this type of regenerator must of necessity have long displacers and cylinders in order that the regenerator may retain a temperature gradient. Ideally the construction of this type of regenerator should allow for airflow on both sides of the regenerator surface. As far as model engines are concerned, where weighty displacers are a determining factor, the perforated steel sheet or the fine steel mesh along the annular gap seems to be a more practical expedient.

**Displacer regenerators:** can be classified as those which are constructed wholly or partly from a matrix capable of absorbing and releasing heat. Displacer regenerators can broadly be classified under the following groups: 1 Displacers which enclose within a thin metal tube a matrix of fine wire mesh with perforated discs at both ends. A similar type of regenerative displacer was used as late as 1895 by A & H Robinson (fig 3.11) who constructed several of these popular 'one-man' or 'two-man' engines, also called 'domestic motors'. 2 Displacers partly filled with regenerative matrix, normally alternating a layer of matrix with horizontal perforated discs, the layers reducing in thickness from the hot end to the cold end. 3 Displacers which are constructed of perforated metal discs or steel gauze discs, either supported by an outer cylinder of gauze, or the discs being mounted on a non-conductive metal or ceramic rod (fig 3.12). In these cases the displacers fit the cylinder easily and without friction while allowing the gas to filter through the regenerators rather than along the cylinder wall. 4 Displacers, made of thin metal, along the outer circumference of which is wound a heat absorbing metal which acts as a regenerator when heated or cooled alternately. One of the finest examples is Stirling's first engine described earlier in this chapter. 5 Displacers constructed from two, three or four light-weight, air-tight cylinders, mechanically joined together. By this arrangement the various sections of the regenerator are separated and therefore the conduction of heat along the regenerator is interrupted.

The adaptability of any of the above regenerative displacers to model Stirling engines is a matter for experimentation and research.

From time to time contributions on the Stirling-cycle/hot air engines appear in *Model Engineer* (published by MAP Ltd) and in *Engineering in Miniature* (Tee Publishing Ltd). Usually these are written by model engineers who have become fascinated by the Stirling engine and who spend a great deal of their spare time developing increasingly efficient model engines.

**Left** Fig 3.10 *Andy Ross' sleeve regenerator used in a 35cc Rider Stirling engine.*

**Right** Fig 3.11 *Robinson's displacer of 1895.*

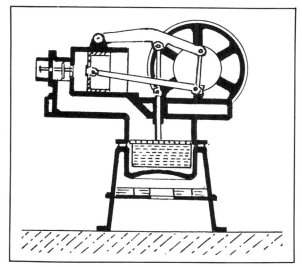

**Below** Fig 3.12 *David Urwick's experimental gauze disc displacers.*

One such contributor made an in-depth study of the regenerative displacer. Mr David Urwick has written and published a number of very informative articles on various aspects of the Stirling engine. One of the more detailed and time consuming studies was that on regenerators. Mr Urwick's experiment centred on the use of stainless steel wire gauze discs (30/40 mesh), enclosed within a wire gauze cylinder, the whole assembly being used as a regenerator/displacer. Careful note was taken of the number of discs in relation to the performance of the testbed Stirling engine, any alteration to the position and number of discs being recorded. These articles appeared in *Model Engineer* of February 7 and 21 1975, and are highly recommended reading. Two other

Fig 3.13 *David Urwick's experimental engine.*

contributors, Mr R.S. Robbins and Mr Andy Ross, in their contributions of June, July and August 1981 gave details of a 'sleeve' regenerator. In both cases steel shim/stainless foil, .0001/.0002 in was used, with dimples tapped or embossed with a tracing wheel. Both contributions are worth reading. Quite apart from the relevant part on the regenerator, the description of the engines provides excellent know-how.

Another experiment, on-going at the time of publication, involves an external regenerator of the first type mentioned earlier in this chapter. The engine consists of a two piston concentric arrangement (displacer and power pistons within one cylinder), with a heater, regenerator and cooler in an adjoining cylinder. The gas leaves the cylinder, goes into the regenerator assembly and re-enters between the power piston and the hot piston. The regenerator is made up of some 200 wafer-thin iron strips (1¼in long, ¼in wide and ¹/₆₄in thick), inserted in the direction of the airflow. The aims of these experiments are twofold: to prove that a model engine can perform satisfactorily with this type of external regenerator and secondly to find and test the best type of regenerative material for such a small scale regenerator. The regenerator body in this experimental engine is constructed in such a way as to allow for changes of the matrix without too much trouble. In order to obtain accurate readings the heating and cooling elements are common to all experiments; heating by an electric element wound on the hot chamber, and cooling by water conduction in the cold chamber.

## DESIRABLE CHARACTERISTICS OF THE REGENERATIVE MATRIX

The size and type of the regenerative matrix depends on the requirements of the regenerator vis-a-vis the total efficiency expected from the model engine or from a small hot air engine. It is therefore essential to remember the main characteristics of the regenerative matrix. Broadly, the types of matrix can be divided into four general classes. **The solid large matrix** which allows maximum heat capacity. This improves the effectiveness of the regenerator while retaining the maximum ratio of heat compared to that of the gas which comes in contact with it. **The large finely-divided matrix** which gives maximum heat transfer by allowing the gas to pass in contact with the largest surface area. **The small porous matrix** which minimises flow losses, thus saving in energy expended. **The small dense matrix** which minimises dead space and improves the ratio of maximum to minimum pressure.

The above characteristics tend to be conflicting and therefore the choice of the right type of matrix is to a large extent guesswork unless, by means of experiments with a removable regenerator casing or body, the various types of matrix can be alternated and the results plotted and compared.

One factor that has a direct effect on the efficiency of the regenerator is gas turbulence. Empirical studies indicate that a turbulent gas flow, where the gas is made to flow in an uneven and irregular fashion in its passage through the matrix, gives a higher rate of heat transfer. In turn this releases and extracts greater efficiency from the regenerative matrix. Robert Stirling recognised this factor and in his 1816 patent application described the passages of the economiser as having 'their sides jagged and rough by bodies projecting from them, as represented at Figure 4 . . .', which is reproduced here as fig 3.14.

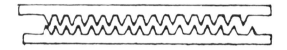

Fig 3.14 *Stirling's first regenerator of 1818—simple sketch.*

Computer-aided research has produced an extensive range of designs for engines and particularly for regenerators. Similar programmes may even be undertaken with small home computers provided that the computer result is corroborated by actual practical design and construction and subject to the availability of the right materials.

In experimenting with regenerators the following basic factors should be borne in mind. We are dealing with 'model' engines and it may be found that the scaling down of regenerators used on larger engines will not produce the same or ratio-reduced efficient results. Secondly there are now more materials suitable for the construction of engine parts and regenerators than before. Therefore prior to designing and building a model engine, it is worthwhile testing interesting materials particularly with respect to their heat absorbtion and conduction characteristics. Materials that could be evaluated include: glass beads and glass rods; ceramic beads and ceramic rods; different types of wire wool/steel wool; fine bore, narrow gauge steel or iron tubes; strips from transformer laminations; fine galvanised wire in short lengths or tight coils; successive thicknesses of wire used as above, and other materials.

Various considerations are taken into account in the design of the regenerator, including the following: size of engine; volume of gas; estimated speed or speed required; regenerative materials available; type of gas to be used — air, hydrogen, helium etc; amount and type of heating available; amount and type of cooling available. Before these parameters are considered, two apparently conflicting requirements are to be kept in mind in the design of the regenerator, the problem of heat transfer and the problem of flow resistance. It is accepted that in order to obtain efficient heat transfer the gas must, in the shortest time possible pass in very close contact with the material of the matrix. In turn this produces flow resistance to the gas. It is therefore important that a balance be obtained between the absolute requirement of good heat transfer and the goal of minimising flow resistance, without impairing the regenerative effect.

Chapter 4

# HEATING AND COOLING

Hot air engines require high external temperature at the hot end and efficient cooling at the other end of the displacer cylinder. Therefore careful consideration should be given to both the heating and cooling systems. To obtain maximum efficiency from a model hot air engine at normal atmospheric pressure, one should aim to obtain a high difference factor, that is heating of up to 800°F and cooling of around 100°F. This may be difficult to achieve under normal home workshop conditions but the nearer one gets to this difference factor of x 8, the more efficient the engine becomes.

## HEATING

Hot air engines are multifuel engines, that is they obtain heat from different fuels — oils, coal, gas, petrol, paraffin, wood etc — any combustible material that gives intense and preferably concentrated heat. In the case of modern engines, where the size of the displacer cylinder determines the size of the flame it is usual to turn to gas burners; methylated spirit burners are often used on the smaller model engine. Heating by an electric element in the hot end of the displacer cylinder (with proper safeguards for safety) is sometimes used on demonstration or research models. Methylated spirit burners may be of a simple wick type or the vaporised burner type, where the spirit is initially turned into combustible gas. Methylated spirit pellets, cubes or bricks may also be used in the same manner as the wick type burner for horizontal cylinders. Obviously the heat generated by these burners is limited and so is the efficiency and the duration of the working power. Such burners are usually used on engines built to demonstrate the principle of the engine rather than to obtain efficiency and power. Surprisingly perhaps to the uninitiated, speeds of over 1,000 rpm can be achieved.

A gas burner, on the other hand, is more efficient for several reasons; it heats a cylinder quickly, it produces far more heat, it is cleaner, gas is more readily available and a burner can be constructed for any size of cylinder in any position. An easy and readily available source of gas heating is the bunsen burner — an efficient way to test most types of engines under workshop conditions. It can be used for most horizontal displacer-cylinder engines except those that have a very low profile. Alternatively a bunsen burner can be

**Left** Fig 4.1 *Multi-wick spirit burner* (C. Vassallo).

**Right** Figs 4.2-4.7 *Types of gas burner.*

used as the mixer tube attachment for most burner devices described in this chapter. The mixer tube can either be pressed in with a tight fit or threaded with a fine thread to fit a corresponding female thread on the burner casing. In either case care must be taken not to allow any gas leakage.

The most elementary type of gas burner is one built like a smoker's pipe: the gas flame emerges at right angles to the gas intake (fig 4.2).

The burner itself may be made in different shapes and sizes to suit the cylinder to be heated. Basically this burner consists of a rectangular box (for example ¾in x ½in x ⅜in high) to which is brazed a ¼in ID tube 3in long. The tube has a ³/₁₆in hole drilled ¾in from the open end, the hole going through both sides of the tube. A gas jet is firmly inserted at the open end, the tip of the jet just about to appear when seen from the side holes (fig 4.3). The combustion box is covered by a metal plate (¹/₁₆in thick), which is then drilled by some 30 holes of ¹/₃₂in diameter. The gas flame may be concentrated by shaping the drilled metal plate in a concave shape (fig 4.4), or slightly spread by shaping the plate in a convex shape (fig 4.5). The width of the flame can also be regulated by the shape of the box — that is by increasing the length to width ratio (fig 4.6). The flame can also be made 'fan' shape by constructing the flame box in almost the same shape (fig 4.7). A fan-shaped flame gives a better heating surface since the flame covers a larger area of the circumference.

The efficiency of these burners depends on the volume of gas to air mixture and on the number of gas holes. Initially one should start with a small number of holes, say twenty, and an oxygen vent of ³/₁₆in. If the gas flame is yellow or yellowish, or if the flame is clinging, the oxygen vent holes should be slightly enlarged, one at a time. If need be, one or more additional air vents may be drilled. The flame should be bright intense blue with a slight hissing noise. Care should be taken to ensure that the gas jet fitted to the mixer tube is suitable for the type of gas used; LPG and butane gases require

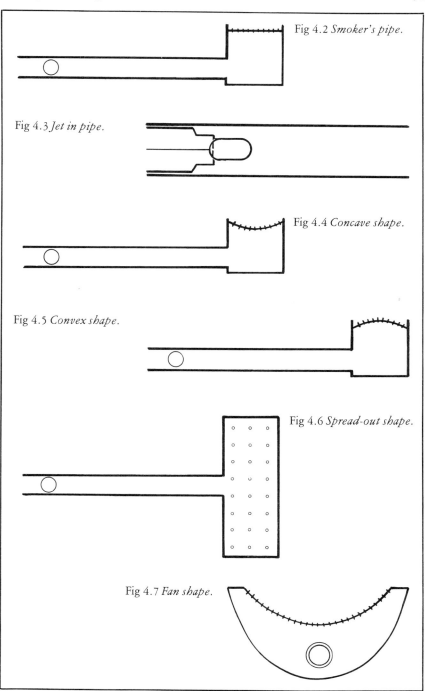

Fig 4.2 *Smoker's pipe.*

Fig 4.3 *Jet in pipe.*

Fig 4.4 *Concave shape.*

Fig 4.5 *Convex shape.*

Fig 4.6 *Spread-out shape.*

Fig 4.7 *Fan shape.*

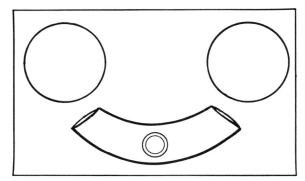

Fig 4.8 *Twin cylinder burner.*

Fig 4.9 *Loop gas pipe burner* (C. Vassallo).

different types of jets. Commercial gas jets can be purchased to fit mixer tubes of most sizes.

A variation of the above burner is the twin-pipe effect whereby two horizontal displacer cylinders may be heated simultaneously (fig 4.8). This burner may be made out of a ½in copper elbow to which a mixer tube is brazed. A quantity of steel mesh or coarse steel wool is inserted two-thirds of the length of the U-tube to prevent lighting back of gas. The tips of the outlets, if flattened slightly, give narrow fan-shaped flames.

Vertical displacer cylinders require different kinds of burners. An elementary type is one constructed from a ¼in ID thin gauge copper (or brass) tube which is bent to form a loop or ring round the displacer cylinder leaving a gap ⅜in all round the circumference (fig 4.9). The tube is segmented by a fine saw or drilled by ¹/₁₆in holes. The average number of segments or holes is 12, but this depends to a large extent on the circumference of the displacer cylinder. Thus if a 1 ³/₁₆in cylinder is to be heated, the inside length of the curved burner tube is about 5½in long; a cut or hole every ½in will give good heating coverage. In addition to the loop or open circle of tubing, an extra length of 3in for the mixer tube is necessary with the oxygen vent holes some ¾in from the open end.

A far more efficient gas burner is one which gives concentrated heat while being economical in the use of gas. This type of burner can be used on horizontal and vertical engines. This burner consists of a metal box which fits around the displacer cylinder; a number of holes are drilled in the internal wall of the box from which jets of intense blue flames are concentrated on the hot end of the displacer cylinder. The 'box' is constructed in such a way as to prevent the escape of gas from any source except the drilled holes. An inner ring or wall is drilled with a number of ¹/₁₆in holes; an outer ring or wall with a space clearance between it and the inner wall of about ⅜in is drilled to take an ⁵/₁₆in ID tube in the centre of its height. Two flat circular discs (washer type) are prepared. The outside diameter is that of the outer wall, while the inside diameter is that of the inside wall (fig 4.10). The inside wall is brazed to one disc first, then the outer wall is brazed. The second disc is then brazed on. Finally the mixer tube is brazed to the outer wall. Care should be taken on all joints to prevent gas escape. Alternatively the discs may be bolted tightly together.

A fine mesh screen tied with one or two wire strands round the inside ring prevents 'lighting back' and allows a better mixture of air to gas, at the same time encouraging the pressure to spread around the internal space rather than finding its way out of the gas holes nearest to the mixing-tube.. Assuming the displacer cylinder has an OD of 1 ³/₁₆in, the inner burner wall should be 2in ID, the outer burner wall should be 2¾in ID. The height of the burner depends on the stroke of the engine, but ¾in should normally be sufficient. The mixer tube is ⁵/₁₆in ID and 3in long, with two ¼in air vents ¾in from the open end. Twenty-four gas holes are drilled inside the burner, preferably

Fig 4.10 *A single-ring gas burner* (D. Urwick).

**Above** Fig 4.11 *Left: A twin-gas burner as used on 'Sturdy' (See Chapter 11). Right: An open-type semi-circular burner used on large diameter cylinders.*

**Below** Fig 4.12 *David Urwick's twin-burner.*

in two or three layers, the holes alternating in each row.

A twin-burner version can be built on the same principle, with an oval shaped box enclosing two burner rings (fig 4.11). In this case, however, the mixer tube is ⅜ in ID and brazed mid-way between the burner rings to allow for even distribution of the gas/air mixture. If brazing facilities are not available, the same type of burners can be made by the use of bolts; in these instances the washer-type discs are made wider that the outer burner ring to accommodate the bolts while thick gauge mild steel is used to resist warping under heat or the tension of the bolts (fig 4.12).

A canopy may sometimes be required to cover the burner during competitions or demonstrations. The construction of the canopy requires metal fabrication with brazing facilities or fastening by screws and nuts. Two round plates of equal size are prepared, one for the back and one for the front with a hole cut in the middle only slightly larger than the diameter of the cylinder to be heated. A side plate to go round ⅚ of the circumference is brazed on. The remaining sixth is the bottom part of the canopy from which air is allowed into the canopy to form a draught (fig 4.13).

The side plate is welded to the front plate first. The position of the burner is then clearly marked on the back plate and a hole drilled for the mixer tube. The burner is brazed or bolted to the back plate allowing air to pass on both sides of the burner. The back plate is brazed to the side plate ensuring alignment of the front cylinder aperture to the ring burner aperture. A short chimney stack is welded on top, a large hole or a number of holes being drilled into the top of the canopy. Finally two L-plates are brazed or bolted to the bottom giving sufficient clearance between the canopy and the engine base.

**Below left** Fig 4.13 *Canopy effect.*

**Below right** Fig 4.14 *One of the most successful gas burners built for model engines (Ross kit B-20 Rider engine).*

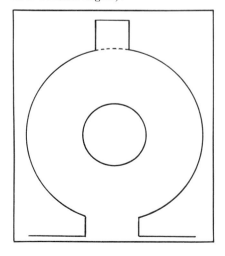

Two excellent burners are described and illustrated by Robin S. Robbins in *Model Engineer* of June 5 1981 and by Andy Ross in *Model Engineer* of August 21 1981. Those who require an even more sophisticated and powerful burner are advised to study the heating arrangement made by Philips for their single cylinder small engines.

## COOLING

The two common methods used for cooling the displacer cylinder are air fin and water cooling, or a combination of both. Further research may yet bring about alternative combinations of these methods and more efficient systems. The basic requirement is for efficient transfer of heat away from the cylinder to the surrounding atmosphere. Therefore close surface contact between the cooler and the cylinder is a prime requirement.

**Air fin cooling:** Small desk-type model engines, used for demonstration purposes and normally run for a short duration, can be efficiently and sufficiently cooled by airfins. Fins are cut from ¹/₁₆in aluminium sheet, and fitted tightly over the cold end of the displacer cylinder with ⅛in brass or copper rings in between to give separation and at the same time good conduction of heat. The size of the fins depends on the size of the cylinder and the method of heating and the temperature at the hot end. A rough calculation may be taken as follows:

Area of fin = 8 x area of cross section of cylinder

No of fins (¹/₁₆in thick ) = $\dfrac{\text{stroke of displacer}}{3}$

Fig 4.15 *Fin coolers.*

Model engines using fin cooling are normally run for short duration but one way of obtaining longer duration without overheating is to fit a small fan, belt driven from the flywheel, sideways to the fins, helping the dissipation of heat by the movement of air through the fins.

**Right** Fig 4.16 *A novel fin cooler* (C. Vassallo).

**Below** Fig 4.17 *Water hopper (right) and water jacket (left) built from household tin containers.*

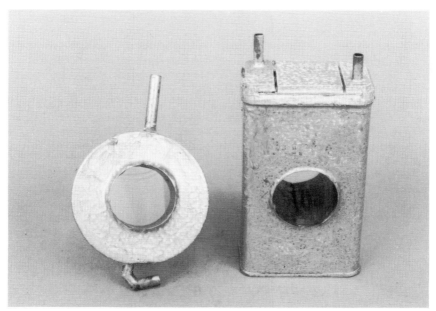

**Left** Fig 4.18 *'Prova I'*
*water jacket.*

**Right** Fig 4.19 *Water*
*tower.*

**Far right** Fig 4.20 *Fan and*
*radiator.*

**Water cooling:** is generally more efficient since the temperature difference can be maintained for fairly long periods. Water cooling allows for higher temperatures to be applied to the hot end. With proper water flow and good close contact the cold end may be kept at a reasonable low temperature for some time.

A simple type of water cooler may be made from a soft lead or copper pipe carefully wound round the displacer cylinder cold end. Care must be taken in this operation. The use of a spring inside the pipe helps to eliminate buckling or distortion, and therefore obstructions. The pipe is connected to a tank with sufficient height to give a fairly fast flow of water. Better still is a water tap.

Another simple type of water cooling system is the hopper type. This can be made from a tin box (Band-Aid, Durkee pepper container or similar). The tin is drilled to fit the displacer cylinder tightly. Once in place the container is sweat soldered carefully to make the tank water tight. A better method is to solder a brass shim (.005in) in the form of a flat ring around the cylinder and then to solder the tin box to the shim. This method allows the water container to be slipped off, while at the same time giving excellent surface contact and heat dissipation through the brass.

Water tanks (fig 4.19) may be made large enough to contain sufficient water for low temperature to be maintained for a reasonably long time. However, the water tower method described below allows for an even longer period of constant low temperature cooling. The water jacket around the cylinder may be made from a small container with inlet and outlet pipes on opposite sides. These pipes are connected by plastic tubing to a long and narrow container (approx 12in x 3in) standing on legs with the bottom level above the level of the water jacket, such that the top pipe of the jacket joins a pipe at the top of the container, while the bottom outlet is joined to the bottom pipe soldered to the bottom part of the container. The container is

filled with water to the level of the upper pipe. When the engine has been running for some time the hot water from the water jacket around the cylinder rises, by thermo-syphon action, to the top of the water tower and is replaced by cold water from the lower end.

Yet another method of cooling the water jacket of a larger type model engine is by using a radiator instead of a water tower, with the same thermo-syphon action as described above, the hot water rising into a multi-spiral or piped radiator, adequately finned. The radiator may be fanned in addition with a mechanical fan from the flywheel (fig 4.20). Such an engine may develop sufficient power to drive a vane-type water pump to obtain a faster circulation of water and therefore not necessarily depend on thermo-syphon action. A typical example of such an engine is the one described in Chapter 15, 'Dyna' which has a radiator, fan and pump all driven by the engine shaft.

**Cooling by highly volatile liquids:** A cooling system using a highly volatile hydro-carbon liquid, pentane is a distinct possibility. Pentane, has a boiling point of 97°F. As a cooling agent it is therefore admirably suitable, particularly in a closed cycle device. If the device is made to consist of a radiator and a coiled tube filled with pentane wound or placed round that part of the cylinder which requires cooling, the action that takes place is as follows. When the liquid heats up to 97°F, the pentane surrounding the cylinder reaches its boiling point, vaporises and rises to the radiator where it cools, condenses and flows back to the cooling tube or coil. As long as the cycle is closed, the action is continuous in its movement. This method was used in the thermo-mechanical generators (TMG) under development at the AERA Harwell but was abandoned after problems from leaking in flexible joints.

CHAPTER 5

# PRESSURISATION

Robert Stirling was the first hot air engine developer to use elevated pressure as a means of obtaining greater engine power, but historians and writers on this subject do not agree which of the engines was the first to be run at 'elevated pressure'. An encyclopaedia published before the end of that century gave the 1816 engine as being the first to work with an elevated pressure but there is no mention of this in the patent specifications. The specifications for the patent granted to the Stirling brothers in 1827 mention a 'vessel containing condensed air', which could be taken as a pressurised reservoir. However a more accurate description is given in the patent specification of 1840 which describes a pressure of air in the air vessels of 150 psi.

Some of the hot air engines built after Stirling's days employed a pump to replace loss by leakage rather than to elevate the working gas pressure, but information about this method is very scanty and cannot be relied upon to give any firm indication of how engine developers of those days felt about the pressurisation of engines. Some writers of those days do mention the fact that 'great energy was expended to supply air at an elevated pressure that very little power was left over for other work!'. As it happened the Stirling brothers were again several decades ahead of their time in perception, when in 1824 they converted a steam engine to run in as a double-acting hot air engine with a force pump to induce elevated pressure. The third engine, started in March 1843, also had this elevated pressure except that in this engine it was preplanned in the design state.

Sealing against losses from escaping gas was a great problem in those days, due mostly to lack of suitable materials. All glands near heated parts were generally sealed by wire, strips or rings of soft lead tightened firmly between two flat surfaces while leather washers soaked or immersed in oil were used around sliding surfaces or rods. Another problem which faced engineers was the lack of heat-resisting metals for the construction of the hot cylinder. It was therefore impossible to raise the temperature of the cylinder to a high degree or to subject it to any high pressure.

A brief mention should be made here about hot air engines which employed the open-cycle (as opposed to the Stirling engines which are of a

Fig 5.1 *Stirling's first ele-vated pressure engine of 1827.*

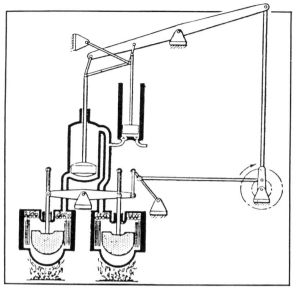

closed-cycle). Most of these engines were made to work at atmospheric pressure. Inevitably these engines suffered loss of working gas through leakage which meant that the engines worked with a pressure which was alternately above and below atmospheric pressure. In order to compensate for the failing, many engines were designed to draw upon injections of air, by means of a one-way (or non-return) valve which allowed air to be drawn into the main cylinder when the pressure fell below atmospheric, or through a small pump working mechanically in conjunction with a springloaded valve and eccentric — or cam-driven from the engine crankshaft, which injected a small volume of air with every stroke. This latter pressure rarely exceeded 20 psi and can hardly be quoted as pressurisation.

The valve system used by these engineers, which for a time included Ericsson, placed these valves at the hot end, which is exactly where they should be so that the charge of air expands at constant temperature. However these valves burnt out with regular monotony due to the lack of heat-resisting metal available. Many modifications were tried, including water cooling jackets around the valves or valve areas. Other designers placed the valves away from the hot cylinder but this was not as successful as Ericsson's later and more advanced models, of which one is preserved in The Science Museum, London. The automatic valve system has come into use again in some model engines built in the last few years and is sometimes called the 'snifter' valve. The Philips air engine, apart from having re-awakened the Stirling-cycle engine and given it a new life, embodied from the beginning of its development programme the concept of pure pressurisation.

## THE CONCEPT OF PRESSURISATION
In hot air engines a small expansion ratio normally occurs, the maximum being twice, very rarely 2.5 times, the minimum pressure. This does not

impart any great power to the crankshaft. Philips scientists realised this and quickly included a mode of increasing power by inducing pressurised working gas into the working space. The availability of heat resisting metals (especially as a result of the gas turbine-related technology), made it possible for Philips to increase both heating and pressurised gas in their very first engine; in the case of heating, from 500°F to 1200°F, and the working pressure (mean efective pressure) from below 1 atmosphere to about 14 atmospheres.

The first Philips 'air motor' had two opposed pistons in a common cylinder with a complicated linkage drive which could not fit very well into a crankcase meant for pressurisation; this layout had another disadvantage, in that leakage could occur from both pistons unless they were extremely well fitting. On the other hand it had the advantage of having an arrangement whereby the cold and hot spaces within the cooler, regenerator and heater layout, were simply and efficiently laid out. In the end the mechanical problems of construction forced Philips to change the configuration to the V-type engine which had a drive totally enclosed in a crankcase. This design gave a simplified linkage to the crankcase allowing the engine to be balanced almost completely. The heater, regenerator and cooler were enclosed in a horizontal passage between the top of the two cylinders. This engine had two working pistons which meant that both had to have a good fit without friction or loss of air. The crankcase was pressurised and this helped to decrease air leakage

Fig 5.2 *Philips first pressu- rised single cylinder en- gine* (Philips).

Fig 5.3 *Flow of gas from crankcase to power cylinder in Philips single cylinder engine.*

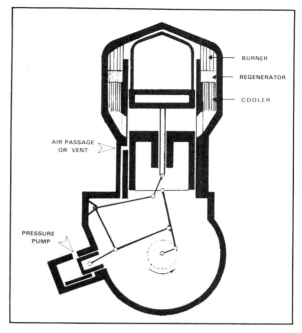

BURNER

REGENERATOR

COOLER

AIR PASSAGE OR VENT

PRESSURE PUMP

from the working area to the crankcase, while also giving substantial assistance to the avoidance of frictional losses.

After considerable research Philips scientists designed and constructed a vertical single cylinder co-axial engine with the displacer (called transfer piston) on top of the working piston. The heater, regenerator and cooler are very conveniently placed in the annular space around the cylinder which meant that the spaces could be kept reasonable small, avoiding dead space while allowing air flow with very little resistance. With this type of layout and construction and with high heat resistance steel and alloys, the Philips engine (fig 5.2 and 5.3) became a winner with a speed of around 2000 rpm and developing several horsepower.

The engine is constructed with a closed crankcase, with only one opening, for the engine shaft (flywheel end). Internally, on one side of the crankcase a small pump directly connected to the crankshaft mechanism, provides air pressure as required, to the crankcase, by pumping air from outside. When the engine has been idle for a long time the pressure in the crankcase drops to atmospheric pressure due to slight leakage which is bound to occur from the engine shaft. When the engine is started there is initially very low power until the pump raises the pressure in the crankcase to the desired level. A connection between the crankcase and the lowest position of the power piston (BDC), allows the flow of pressurised gas from the crankcase to the space between the displacer and the power piston thus equating the minimum pressure of the cycle to that of the crankcase. Therefore the pressure of the latter (the crankcase) of 8 atmospheres brought about by the pump becomes equal to the engine minimum pressure (also of 8 atmospheres). This also means that the power of the engine can be regulated in a relatively simple way. The first engine designed in this manner had a top working pressure of 20 atmospheres (300 psi approximately), a minimum pressure of 8 atmospheres (120 psi approximately) and a crankcase pressure of 8 atmospheres.

The elevated pressure in the crankcase of a hot air engine offers these advantages: 1 Decrease of air leakage from the working space to the crankcase through the piston walls and piston rod (displacer rod); 2 Decrease in frictional losses and 3 lighter load and stress on the bearings and the crankshaft. These advantages are particularly useful in small engines of fractional horsepower and also for model engines of a single cylinder co-axial configuration, although other engines with an enclosed and sealed crankcase can benefit from an elevated pressure.

## SEALING PROBLEMS AND RELATED DEVELOPMENT

In large modern Stirling engines the working gas is usually helium or hydrogen; these gases are installed under very high pressure, in some cases as high as 200 atmospheres (almost 3000 psi). To contain this pressure there is a need for highly effective seals between the working space and the crankcase and around the engine shaft. These large engines also need shaft lubrication in view of the high speed and power output. The dangers of lubrication are the possibility of explosion or the contamination of the regenerative matrix which can also result in a burnt-out regenerator.

One development in the research for effective seals in the working area has been the 'rollsock' seal (fig 5.4) which consists of a polyurethane tubular sleeve, very much like a sock, one end of which is attached to the piston rod and the other to the engine housing. This sleeve is very flexible and forms an effective barrier against escaping gas and oil contamination. Gas pressure on one side (the working side) is counterbalanced by high pressure oil held in a reservoir under the seal. Although this is considered a breakthrough in sealing techniques, the high cost of installing mechanism to provide the balancing oil

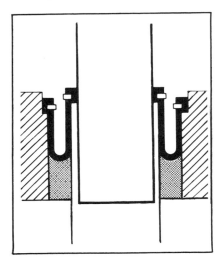

Fig 5.4 *Rollsock seal.*

pressure makes the rollsock seal a very expensive item in the engine construction. The crankcase/engine shaft is another problematic area although a number of different types of seals have been developed, including rotary seals, hydrodynamic seals and many others. The need for seal replacement usually arises when the engines have been running for some time and if the engine requires recharging with expensive gas, as opposed to free air the expense is compounded.

Small engines designed for electricity generating are designed with an inbuilt generator with only the power leads leaving the crankcase, making it relatively easy to provide effective sealing. Such engines usually have a small internal starter motor. This layout brings a Stirling engine/generator with crankcase pressurisation well within the bounds of mass production.

## SIMPLE EXPERIMENTS WITH A MODEL ENGINE

The engine 'Sturdy' described in Chapter 11 has been the subject of some simple experiments in pressurisation. The configuration of the engine does not really lend itself to serious experiments since a co-axial single cylinder layout is a much better engine to pressurise. 'Sturdy' was fairly easily and readily adapted with minor construction work. In the first place the rear end of the engine was sealed off completely after the end part of the shaft had been removed. A ¼ in dural plate was accurately cut to fit into the engine compartment, bonded in place with Super Epoxy and bolted by 6BA screws. The bonding was checked afterwards for cracks or pin holes. A bicycle tyre valve was fitted and bonded to the side of the engine compartment between the flywheel and the power cylinder on the left-hand side when looking at the flywheel from the front. The engine compartment required the drilling of a hole and some filing so that the valve could be accurately fitted from inside with fine rubber (ex-football bladder) washers on both sides of the engine compartment wall. Bonding is not really necessary, but since the engine was being completely and (hopefully) permanently sealed airtight, no chances were taken for minute leakage points in that assembly.

Finally, the front end was sealed off. The major part of construction work was in this area and it necessitated both planning and precision bench fitting. Two end plates were required, one internally, within the engine compartment and one externally to complete the sealing process. The internal plate, made from ¼ in dural was cut to fit into the engine box very accurately after having drilled a hole for the engine shaft with $1/64$in to spare radially. This plate was pushed as far back as possible against the crankshaft bearing block to allow the biggest possible gap between the internal plate and the external plate and permit the fitting of an adequate rotary seal. The plate was bonded in place with 6BA bolts, screwed in to prevent any movement of this plate during later stages of construction work. Then the external plate was prepared. The size of the plate is that of the outside dimensions of the engine compartment since this plate, acting as a flange, has to seal tight against the compartment's thin walls with either a rubber gasket or bonding material between the compartment and the external plate. This plate is also adapted to take the engine shaft.

Two methods were used for sealing the engine shaft, one method superseding the other. Both are mentioned here although the second one was more successful in retaining engine pressure for a longer time. The first sealing method involved the fitting and bonding of a sealed ball bearing into the external plate with a rotary sealing outfit between the external and internal plates. This assembly consisted of the following: **1** a 1.5 mm 'O' ring of the same diameter as the engine shaft, fitting right against the ball bearing inner ring; **2** a football bladder elastic washer, fitting tightly onto the engine shaft with an external diameter of 1 in; **3** two concentric fine compression rings, one of ¼ in to press the rubber washer against the 'O' ring, the other ½ in to press the rubber washer outer edge against a ball bearing outer ring; **4** a fine washer made of brass shim, ³/₁₆in ID, 1 in OD, resting against the internal plate. The assembly is inserted in reverse order: brass shim washer, inner compression spring, outer compression spring, fine rubber washer, 'O' ring. The external plate is then placed against the engine compartment with sealing or gasket compound and bolted in the four corners to the internal plate.

The engine compartment was tested for airtightness. A small bicycle pump was used to force air into the crankcase through the bicycle valve and in the first test the compartment was placed in paraffin. No air bubbles emerged. The next test was a time-lapse one. The engine was left slightly pressurised for some time, and at 15 minute intervals the valve tested for air pressure. Up to one hour later air could be heard hissing when the valve was pressed in. The fault in this seal outfit appeared when the engine was run, with the working air escaping within a few minutes.

The second sealing method followed Mr F. Brian Thomas' pressurised engine seal (fig 5.5) with a PTFE lip seal inserted between two sealed ball bearings in a precision drilled engine shaft guide bonded into the external plate. This seal caused slightly more friction but gave greater 'airtightness' to the engine. The pressurisation experiment was sufficiently successful to warrant mention while further experiments are on the drawing board.

To test the first seal the engine was run at normal atmospheric pressure. While it was running the crankcase was slightly pressurised by a few strokes of the bicycle pump. There was a sudden increase in speed, from 600 rpm to about 800 rpm, which was maintained for about five minutes, after which the engine speed dropped to normal. No hiss could be heard when the valve was pushed in indicating that all excess pressure had escaped. The second test was to pump the air into the crankcase before the engine was started. The engine took much longer to start, but when it did it took off at increased speed, levelling off to normal again within a few minutes. The crankcase was then pumped again, the speed picking up rapidly until the crankcase was overpressurised and the engine speed dropped considerably until stalling occured.

The third test involved running the engine at slightly elevated pressure, with the occasional stroke of the pump every three to five minutes to maintain pressure, taking care not to over-pressurise, while watching out for engine speed. This went on for about one hour when the engine started overheating in spite of continuous water flow through the cooler. This experiment was repeated several times with the same result.

The experiments were repeated with the second seal with one notable difference — the air pressure in the crankcase lasted longer and the pump topping-up was less frequent between strokes, but there was no more increase in engine performance than with the first seal.

The results obtained showed that an elevated pressure of a few pounds in the crankcase made enough difference to warrant more investigations into the effects of pressurisation. It seemed obvious that some of the pressure went through the displacer rod guide bush into the working area. This slightly elevated pressure (not more than 20 psi), increased speed by about 20 per cent, although the torque was appreciable higher. On the other hand engine design and configuration were not conducive to either increased pressure or increased performance.

## THE 'BRIAN THOMAS' SELF-PRESSURISING STIRLING ENGINE

The late Mr F. Brian Thomas built a self-pressurising Stirling engine which he entered for the 1978 Hot Air Engine Competition, winning the award.

The engine is a rhombic drive, water-cooled single cylinder engine which was extended beneath the rhombic mechanism to take the pump body. The pump con-rod is directly linked to the rhombic drive, thus avoiding another link system to the crankshaft. The pump can bring the internal pressure up to 60 psi, but the engine operates best at 40 psi (just under 3 atmospheres). An adjustable safety valve and a pressure gauge are mounted on the engine cover. The description of the air pump (air compressor) construction as given by the late Mr F. Brian Thomas is reproduced here by permission of Mrs E. Thomas.

'The pump body or cylinder was made from free-cutting mild steel hexagon rod, $\frac{5}{8}$ in AF and the bore was drilled and reamed $\frac{5}{8}$ in. The air inlet port is $\frac{1}{32}$ in and is just cleared by the piston at TDC. The piston itself is of $\frac{1}{8}$ in ground silver steel rod which is drilled as shown using a No 52 drill (.0635in). This gives the right clearance for the $\frac{1}{16}$ (ball bearing) ball valve. Keep the $\frac{1}{32}$ in hole at the lower end of the piston as short as possible to cut the clearance volume to the minimum. To get a good seal, the ball is hammered into its seal using a small punch. The very light compression spring [wound by Mr Brian Thomas from spring wire] is .010in diameter and the OD of the spring must be less than $\frac{1}{16}$ in. The top end of the piston carries a 10 BA screw which is drilled to receive a short length of Bowden cable, which is soft soldered into the hole.

The outlet air port is drilled $\frac{1}{32}$ in, just below the lower end of the 10 BA screw. It must be clear of the 'O' ring near the upper end of the piston. The top end of the flexible pump piston connecting rod is soft-soldered into a hole drilled in the head of a 5 BA hexagon headed screw which screws into a threaded hole in the lower end of the displacer rod. The Bowden cable compressor rod is shown in the general arrangement drawing. It is about $\frac{1}{2}$ in long and is adjustable for length at the 5 BA screw with its lock nut. The piston must be adjusted so that its lower end is only about .002in from the pump cap at BDC. The cap is machined from the same hexagon mild steel bar as the pump and is sealed with an 'O' ring. Probably the most vital part of the pump is the 'O' ring which seals the upper end of the piston. This is of white silicon rubber $\frac{1}{8}$ in ID. Until this ring was fitted the pump never produced more than 25 psi. As soon as the ring was fitted 60 psi was obtained.

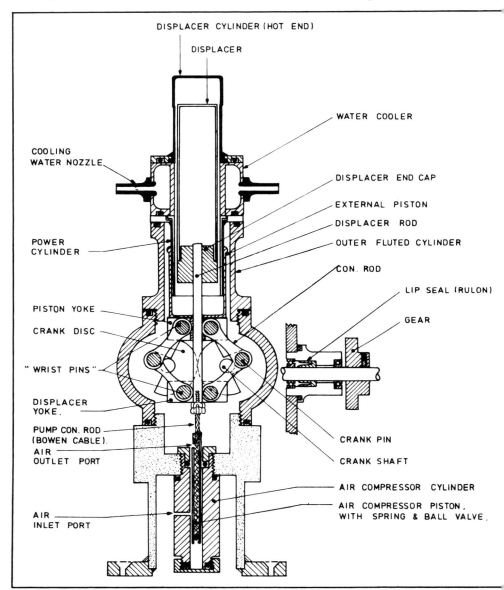

Fig 5.5 *Brian Thomas' self-pressurising Stirling engine.*

When the completed engine is finally assembled it must be tested for leaks as even the smallest leak will stop any worthwhile pressure building up. The adjustable safety valve carries a cycle tyre valve so that the engine can be pumped by hand, any major leak will show immediately by a rapidly falling pressure on the gauge, and when hand pumped the engine should hold its pressure for several hours. If it does not, the leak must be found. The only way to do this is to pump the engine up to 60 psi and then submerge it in paraffin. After curing all final leaks, the engine must be very carefully cleaned to remove all paraffin and the outer pair of ball races re-oiled.'

Mr F. Brian Thomas' contribution in *Model Engineer*, May 19 1978, also includes a few notes on the construction of the engine in addition to some engine data.

Pressurising an engine is no doubt a fascinating development but it should not be the end-all of the humble Stirling-cycle engine. Much can be learnt from Andy Ross' recent experimental engine which developed 44.1 watts at 2,750 rpm without pressurisation!

CHAPTER 6

# DESIGNING AND BUILDING MODEL STIRLING ENGINES

*It is assumed that the modeller has built at least one of the engines described in Chapters 9 and 10 and that he has grasped the principles of how the engine works before he decides to build an engine to his own design.*

A study made by R. Sier and published in the *Model Engineer* (September 3 1976), found that the design of the first Stirling engine followed broadly these parameters:

**1** The length of the displacer chamber = 3 times its diameter.
**2** The length of the heater chamber = $2/3$ of the length of the displacer chamber.
**3** The length of the cooling chamber = $1/3$ of the length of the displacer chamber.
**4** Swept volume of the displacer = $1\frac{1}{2}$ times the swept volume of the power piston.
**5** Length of the displacer = $2/3$ of the length of the displacer chamber.
**6** Stroke of the displacer = $1/3$ of the length of the displacer chamber.

Indeed, from calculations made by the author of other engines by Bailey, Lehmann and many others, these parameters were almost always followed. This says much for Stirling's perception and grasp of the thermodynamic principles at that time.

Not all these parameters can be followed in model engines. For example; to have a heater chamber equal to two-thirds of the length of the displacer chamber (cylinder) requires a super cooling method not easily obtained in a small engine. Also, in Stirling's days the metal used was poor in quality and massive in size accounting for the need to heat a large area. With model engines constructed of bright mild steel, heating the rear third of the displacer cylinder is sufficient. Moreover if conduction of heat along the cylinder wall is avoided, better regeneration from the displacer is obtained and cooling is greatly facilitated. Stirling's displacer was two-thirds the length of the displacer chamber. This method is used when the displacer serves also as a regenerator, as in most model engines. When the regenerator is contained in a separate body, the displacer length can be shortened substantially, as in the Philips' air motor. Some of the parameters are still generally followed in the construction of simple model engines. These are: compression ratio; displacer cylinder length; displacer stroke and cooler area.

## GENERAL NOTES ON ENGINE DESIGN

**Engine configuration:** Three basic configurations are described in this book. The projects detailed in Part 2 are based on these layouts, starting with a simple engine and working progressively towards more sophisticated and interesting models. The three basic configurations are: **1** Twin cylinders (power and displacer) in parallel formation; **2** Twin cylinders in V-formation or at 90° to each other; **3** Single-cylinder, co-axial with the power piston and displacer in the same cylinder.

A twin-cylinder engine, such as 'Dolly' in Chapter 9, has the two cylinders parallel but on opposite sides of the cylinder plate. Another version may have the parallel cylinders on the same side of the cylinder plate with the interconnecting air passage leading from the centre of the displacer cylinder to the top of the power piston. The V-formation twin-cylinder engines, such as 'Lolly' in Chapter 10 and 'Sturdy' in Chapter 11, have the power cylinder at right angles to the displacer cylinder, both con-rods working off the same crank-pin or crankshaft. The co-axial, engines, the most efficient type of all, such as the Ericsson engine in Chapter 12, 'Prova II' in Chapter 13, 'Sunspot' in Chapter 14 and 'Dyna' in Chapter 15, have the displacer and the power piston in the same cylinder. These have a single crankshaft but a double-throw crank to give different strokes and the required phase angle. The only difference amongst the above four engines is the method of regeneration in 'Prova II'. There are other engine configurations, some of which may be dealt with in future publications.

**The twin-cylinder parallel configuration** requires a simple and easy running mechanism capable of providing for alterations to phase and stroke without

Fig 6.1 *Parallel cylinders with double flywheel mechanism* (C. Vassallo).

complications. Two types of drive, the flywheel/disc drive and the double crank mechanism, are readily adaptable for this configuration.

The flywheel/disc drive mechanism (fig 6.1) is really a combination of a flywheel and a disc (or web), or two flywheels. The assembly is mounted on a pillar or block, such that the flywheel on one side of the block is connected to the power piston, while the flywheel, disc, or web on the other side of the block is connected to the displacer. The block is sufficiently wide to keep the shaft from twisting. Alternatively two pillars are used with a brass bush at each end. The advantages of this type of drive are twofold. The phase angle can be varied — simply by altering the positions of the flywheel and web relative to each other (instead of being at 90° to each other, one flywheel or disc can be turned to give say 95° or 85°). Secondly, the stroke of the power piston or the displacer can be lengthened or shortened by simple repositioning of the crankpin.

The double crank mechanism for a parallel cylinder engine is slightly more complex (fig 6.2). The crankshaft passes through a block or pillar placed to one side of the cylinder assembly. The crankshaft has a disc or web with a crank pin designed to give the right length of stroke to the near-side cylinder, normally housing the power piston. The crank pin is made sufficiently long to take on another web with its own pin set at 90°C to the first crank, and giving the necessary stroke length to the displacer on the far side. Obviously the crank pins should be kept as short as possible to avoid too much stress on the webs. This in turn means that the cylinders should be close to each other, one almost behind the other. This configuration also helps to keep the connecting air passage short, thereby lessening dead volume.

**The V-engine** is a fairly easy type of engine to design, needing only some construction practice. It has the advantage of a compact and sturdy mechanism. V-type engines may be designed to have the displacer cylinder horizontal, with the power cylinder vertical; the power cylinder horizontal with the displacer vertical or a perfect V formation with both cylinders at 45° from the horizontal.

Since a single crank web and pin is used to drive both con-rods, it is obvious that to obtain the ratio of 1:1.5, two different sizes of cylinders are used, a wider diameter for the displacer cylinder and a narrower diameter for the power cylinder. And slight variation of the ratio will not necessarily effect the engine performance. The positioning of the cylinders is not critical, and to a large extent depends on the burner/cooler facilities available. Thus if a single bunsen flame is used, the displacer cylinder is best placed in a horizontal position. If a ring burner is used, any of the three positions is suitable. If the displacer is slightly heavy, the displacer cylinder is best in a horizontal position. The V-type configuration (fig 6.3) works best when the weight of the power piston and the displacer are almost equal. In designing this type of engine, one has to keep in mind that the length of the con-rod from the displacer to the crank pin is always longer than the power piston con-rod and therefore the displacer cylinder is placed further away from engine frame, the cylinder being supported either by distant pieces or on pillars.

**The single cylinder co-axial or concentric engine** (fig 6.4) is by far the most

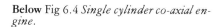

**Below left** Fig 6.2 *Parallel cylinders with a double crank mechanism.*

**Right** Fig 6.3 *The mechanical drive of a V-type engine.*

**Below** Fig 6.4 *Single cylinder co-axial engine.*

efficient type of engine both in design and performance. Almost all engines being designed and constructed today are of this configuration. The main reason for the efficiency lies in the fact that in a compact engine of this type there is very little dead space or volume between the displacer and the power piston. The working gas is contained between the displacer and the power piston, with less chance of escape and the whole volume of gas is utilised in providing energy. The difficulty in building this engine lies mostly in the design and calculation stages, as it is difficult to visualise the various movements inside the cylinder during the various stages of rotation. There are practical difficulties to be overcome in the construction of single cylinder co-axial engines — the displacer rod has to pass through the crown of the power piston and down through either a hollow piston rod or through an internal guide bush to reach the crankshaft. Sealing problems have to be overcome — the obvious points of leakage are the power piston wall and the displacer rod exit. Other practical difficulties are that the power piston tends to overheat, creating an additional problem of lubrication. Therefore, in designing this type of engine, it should be ascertained that the lower part of the cylinder and the power piston are effectively cooled to avoid heat creep from the hot end to the cold end of the cylinder.

## DRIVE MECHANISMS

One of the more fascinating and challenging aspects of designing and constructing model Stirling engines is that of devising the mechanical drive of an engine. Variations of drive mechanisms are almost endless and scope for inventiveness is unlimited. Indeed many an efficient and interesting method has evolved from the world of model engines. The drive mechanism has to transfer energy from the power cylinder to the flywheel; to provide separate movements to the power piston and to the displacer; to provide different movements to the power piston and the displacer if this is not provided for by the positioning of the power and displacer cylinders (ie, by a V-formation or a 90° assembly); and to provide as near friction-free movement of the crankshaft as possible.

The different types of drive mechanism may be divided as follows:
1 Crank and lever linkage mechanism.
2 Single-throw crank mechanism (V-type engine).
3 Cam drive mechanism.
4 Scotch yoke mechanism.
5 Discs/webs mechanism.
6 Gear drive mechanism.
7 Swash-plate or wobble-plate mechanism.
8 Siemans drive mechanism.
9 Nutator drive mechanism.
10 Rhombic drive mechanism.
11 Ross linkage mechanism.
12 Novel drive mechanisms.

**The crank and lever linkage mechanism** is a complicated method of driving a model engine, involving stress and friction on the different components. This system was widely used in olden days because no other system had been

Fig 6.5 *Crank and levers linkage mechanism.*

Fig 6.6 *V-type single throw mechanism* (C. Vassallo).

devised and because the developers of that age followed (too closely) the design of the steam engine. This mechanism does not allow an engine to develop high speed, although it is delightful to watch. Power output in relation to energy used to heat the engine is very low.

**The single-throw crank mechanism** (fig 6.6) used in the V-type, on the other hand, is more efficient and capable of sustained stress and high speed. Some of the high powered Stirling engines currently being developed have this configuration, with four or six cylinders in line on one single crankshaft. Modern engines using this mechanism are generally very efficient and can reach quite high speeds with appreciable power.

**The cam-drive mechanism** (fig 6.7) is rarely used since it requires careful machining and perfect alignment, but once the machining operation is successful, an engine can achieve quite high speeds. The cam-drive is usually used on the displacer con-rod while the power piston is driven by a web-con-rod mechanism. The only old engine known to have this type of mechanism was produced by Schwartzkopff in Berlin around 1860. Nothing much is known about its efficiency, but the mechanical operation was quite successfully copied on a model engine. The cam was designed to raise the displacer during 120° of crankshaft rotation, dwell for 60°, lower the displacer during the next 120°, dwelling again for another 60°.

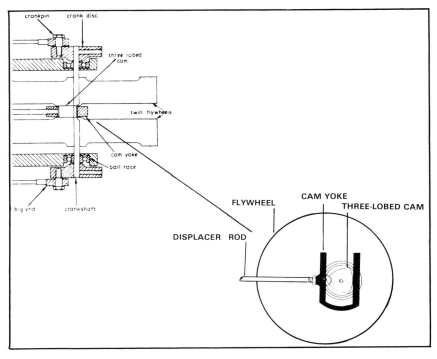

Fig 6.7 *Cam-drive mechanism* (Brian Thomas).

The late Mr Brian Thomas described his 1,500 rpm model engine in *Model Engineer* of October 3 1975 using the cam illustrated in fig 6.7.

**The Scotch yoke mechanism** (fig 6.8) has been used quite successfully on several model engines — this drive is generally used where space or design requirements do not allow for long con-rod stroke. Displacers can have long con-rods shortened by the use of the Scotch yoke. The illustration (fig 6.8) shows an example of this mechanism used on a displacer rod (with an ordinary con-rod attachment for the power piston on the far side). This mechanism is favoured by modellers who require compact engines for applications with restricted engine space. One of these engines, with the mechanism enclosed in a perspex cover, was specifically designed as a miniature engine, while the other, designed for marine use, has two Scotch yokes on the extreme ends of the engine. The latter engine has an amazing turn of speed and power for its size and can easily move an 18 in-24 in boat at a good speed.

**The discs/webs mechanism** (fig 6.9) is a straightforward assembly which is simple and effective, capable of high revolutions, while being relatively easy to adjust for stroke and phase angle. The disc can be the flywheel to which a crank-pin is fitted. This mechanism is mostly suitable for parallel configuration engines. It is ideal for model desk-top engines or experimental models.

**Above** Fig 6.8 *Scotch yoke drive mechanism* (C. Vassallo).

**Right** Fig 6.9 *Flywheel (disc) and web drive mechanism* (C. Vassallo).

**Below** Fig 6.10 *Gear drive mechanism.*

The **gear drive mechanism** (fig 6.10) is rarely used in model engines although such engines can develop high torque and low speed. Model engines described in *Model Engineer* around 1973 had a bevel-gear mechanism on a single crankshaft to drive both the power piston and the displacer. The illustration (fig 6.10) shows a multiple gear assembly used very successfully by the author on a twin cylinder horizontal marine engine. The bevel gears drive two horizontal displacers while the lower spur gear drives two diametrically opposed power pistons on a fabricated crankshaft.

**The swash-plate or wobble-plate mechanism** is a fairly recent innovation to the model engine field. The first concept of this type of drive was patented by Sir William Siemans (though there is no known working engine described in any literature of the period, circa 1863). The concept was modified into a swash-plate or wobble-plate drive and a number of full scale engines were built with this mechanism. The drive consists of a rotating inclined disc with a crankshaft causing pistons and displacers to move forwards and backwards as the disc rotates. An example of this drive is in fact the swash-plate mechanism illustrated in fig 6.11. This swash-plate mechanism is incorporated in an engine built by Mr Albert Debono of Malta and described by him in *Model Engineer* of September 19 1975. The swash-plate is fixed to the crankshaft with a 'follower' on top to pull and push the displacer rod. The 'follower' holds revolving steel balls to reduce friction. Altogether it is a lively mechanism, delightful to watch and to experiment with.

**The Siemans drive** (fig 6.12) is very often confused with the swash-plate or wobble-plate because of the similarity of movement of the plate or disc, a similarity that is apparent on paper but not in action. This mechanism is best described in Mr David Urwick's own words for *Model Engineer:* 'The Siemans drive plate rocks on a fulcrum and serves as a bell crank, or series of bell cranks, transferring the piston rod thrusts to a single crank pin driving a flywheel or disc. The Siemans pin drive is a positive direct linkage and the throw can be as large or small as desired, like a normal crankshaft. This pin is headed by a ball entering freely into a hole in the crank-disc, drilled at a

Fig 6.11 *Swash plate drive mechanism* (A.N. Debono).

Fig 6.12 *Siemans drive mechanism* (D. Urwick).

radius to give any desired throw. There is no end thrust at all on the main shaft, reducing vibration and friction.' This drive was to be modified by Mr Urwick, replacing the ball fulcrum on which the rocking plate rocks by a Carden-style universal joint.

The illustration (fig 6.12) shows Mr Urwick's Siemans drive mechanism which is described in detail in two contributions he made in *Model Engineer* of December 1980/January 1981 and explained further in the same publication in June 1982. The author had the opportunity to see this lovely engine in action and carries a lasting impression of a very beautiful and precise piece of engineering. While running it is perfectly quiet, smooth and vibration free. A sample test run brought a speed in excess of 1400 rpm in a relatively short time.

A modification of the above drives is the **nutator mechanism** (fig 6.13), which is best described as a multiple ball and socket arrangement the effect of which is a 'nodding' action — hence the word 'nutator'. The mechanism works extremely smoothly when a brief flick of the flywheel starts the swivelling and nodding action of the drive balanced on the various balls and sockets, pushing and pulling the power piston and the displacer alternately while driving the flywheel through the crank half plate.

**Left** Fig 6.13 *Nutator drive mechanism* (D. Urwick).

**Below** Fig 6.14 *Working plans for the Nutator engine* (D. Urwick).

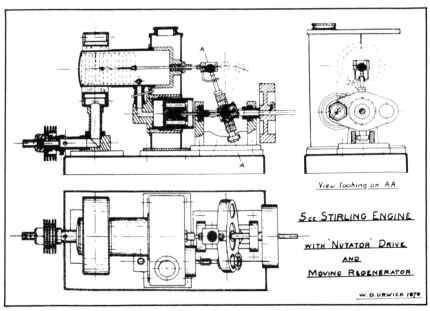

**The rhombic drive mechanism** invented by Philips of Eindhoven, Netherlands in the early days of their development of the Stirling engine (Philips air motor), is a magnificent piece of engineering. This mechanism gives a single cylinder engine several advantages over similar sized engines with different drive mechanisms — the engine can attain and hold high speeds without any lateral or side thrust on the power piston or displacer rod,

**Above** Fig 6.15 *Rhombic drive mechanism* (Prof. C.J. Camilleri).

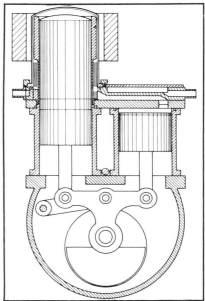

**Right** Fig 6.16 *Ross linkage mechanism devised by Andy Ross.*

and is above all totally vibration free. As a result of its particular mechanism wear and tear, normally minimal in hot air engines, is negligible with this drive. It is also relatively easy to contain this mechanism in an enclosed crankcase with only one outlet, the crankshaft bearing. It takes an experienced modeller with substantial workshop equipment to machine the parts with great accuracy. It also needs an experimenter to have a mathematical approach

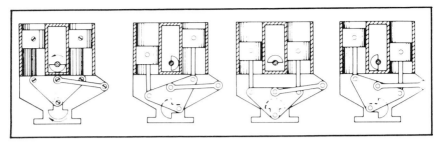

**Above** Fig 6.17 *Ross linkage working movements. The drawings show the positions of the crankshaft and the balance shaft during different parts of the cycle. The gears synchronizing the balance shaft with the crankshaft are not shown, for simplicity.* (Andy Ross).

**Left** Fig 6.18 *Novel drive mechanism* (Prof C.J. Camilleri).

to obtain the right stroke, phase and ratio, while maintaining the efficiency that this engine calls for. The rhombic drive mechanism shown in fig 6.15 was constructed by Professor Charles Camilleri of Malta, and during a trial run in the presence of the author achieved a speed in excess of 1,000 rpm in a few seconds.

**The Ross linkage mechanism** (fig 6.17) is one of the latest to be devised in the model or small engine field. This mechanism consists of a balanced crankshaft with a triangular yoke drive or linkage (called the Ross linkage after its inventor, Mr Andy Ross of USA, Patent No 4,138,897). The yoke is held captive to one side of the crankcase, secured in the centre to a revolving bell crank and linked to a power piston (cold side) at one end, and another power piston with an extension (hot side) at the other end. The engine is in fact a Rider, on which Mr Ross has become an authority. This type of mechanism is slightly complex for the beginner to hot air engine modelling, but it is mentioned here to show the beginner that there is scope for experimentation and development and there is always something new and exciting to achieve.

The 35 cc Rider Stirling engine which first incorporated this linkage is fully described in *Model Engineer,* July/September 1981, while development of Stirling engines by Mr A. Ross himself is discussed in a detailed article in *Live Steam* (USA) of January 1983. (Copies of this article and other Ross literature can be obtained from Ross Experimental Inc.)

Fig 6.18 shows a **novel drive mechanism** constructed by Professor Charles

Camilleri of Malta, for which there seems to be no short descriptive title. The mechanism consists of a double crank machined from a 1in solid mild steel bar on a crankshaft, the crank pins being machined on opposite sides of the webs to give the necessary simultaneous throw to two con-rods. Two rocker arms on either side of the assembly join the lower con-rods and the upper con-rods form the power pistons. When the drive mechanism is activated, the cranks push the lower connecting rods outwards, forcing the rocker arms to push the upper con-rods inwards to perform the compression stage. Meanwhile a crank-pin on the outer web takes a Scotch yoke at the end of the displacer rod, the action of which is at 90° to the crank throw/rocker arm movement. This is a beautifully machined mechanism delightful to watch. It has a no-load speed in excess of 1,000 rpm, picking up revolutions very quickly.

At some stage a hot air engine enthusiast will attempt to design his own engine. Almost all modellers of Stirling engines known to the author have constructed at least one engine to their own design while others have taken this hobby even more seriously and experimented with many different configurations and mechanisms. The following notes are pointers to the requirements of the initial stages of design. As in everything else, the first attempt is bound to encounter problems and difficulties — worse still if the completed engine fails to go after the first few attempts. The reaction may extend to throwing it and the design paper out of the nearest window!

Generally speaking there are two types of model engine designers: those who first design an engine and then get down to the task of finding materials to use and to machine; the others are those who design an engine based on materials available, obtaining the rest as they go along. The author belongs to the latter class with one proviso. Over the years a fair number of cardboard boxes have been stacked in odd corners of the workshop and garage, containing bits and pieces of materials which may one day prove useful! The difference between these two approaches to design is evident, the first can go more by the book while the latter has to improvise and possibly experiment more with the parameters. Anyway it is a worthwhile stage to reach, and when the model to one's own design works, there is the added pride of a double achievement.

Usually the design stage follows this sort of pattern:
1 Decide on the configuration of the engine;
2 Draw the layout roughly to see whether it is feasible and to find out if dead space or dead volume can be reduced or avoided;
3 Draw the layout to scale;
4 Cut cardboard templates to the original size and scale, pin to a board where appropriate (eg, con-rods, links, cams etc), and check for free uninterrupted movement such as bores wide enough to take con-rods etc;
5 Make working drawings, top view, side elevation, front elevation;
6 Draw the drive mechanism to an enlarged scale to obtain precise measurements.

The next stage is construction. Here again experience has taught that planning the stages of construction carefully, but not necessarily in great detail, helps to avoid mistakes. All the more so when the model is

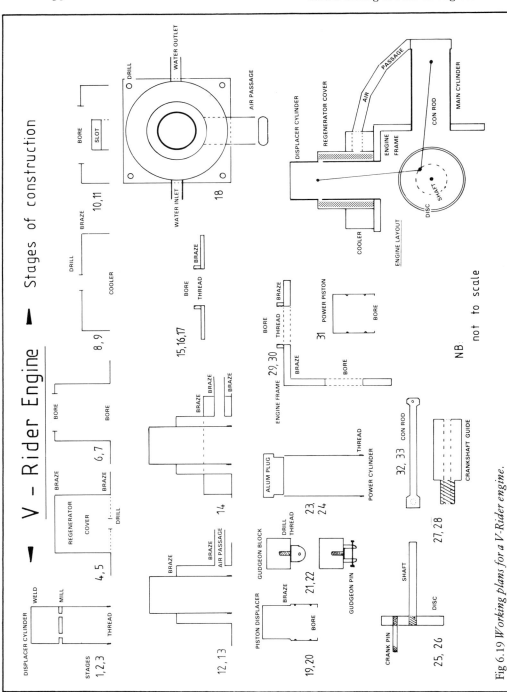

Fig 6.19 *Working plans for a V-Rider engine.*

complicated and involves different machining and bench fitting processes. The drawing for the construction stages should be sufficiently large (board size) to be followed at a glance, crossing out each stage when completed. The illustration at fig 6.20 is an actual construction plan of a V-Rider Stirling engine. In putting the stages of construction on paper before initiating the operation, the modeller is creating a system or method of approach; making a mental inventory of materials required; planning bench fitting and machining operations in correct order, avoiding repetition and re-use of equipment; and finally, ensuring that no construction step is overlooked.

## DESIGNING A CO-AXIAL ENGINE

The co-axial engine requires a slightly different approach in the design stages in view of the position of the power piston and displacer in the same cylinder and the problem of following the movement of the two components during the various stages of the drive. The design stages follow this pattern:

**1** Decide on the following parameters: the ratio and length of strokes of the power piston and displacer; the length of the displacer; the length of the power piston; and the distance of the crankshaft centre from the cylinder-plate keeping in mind that short connecting rods tend to cause lateral thrust.

**2** Decide on the scale of the first engine layout drawing, ie, × 2, × 3, etc.

**3** Draw on the diagram the position of the main cylinder and locate the centre-point of the crankshaft.

Fig 6.20 *Stirling-cycle co-axial engine displacer and power piston movement plan.*

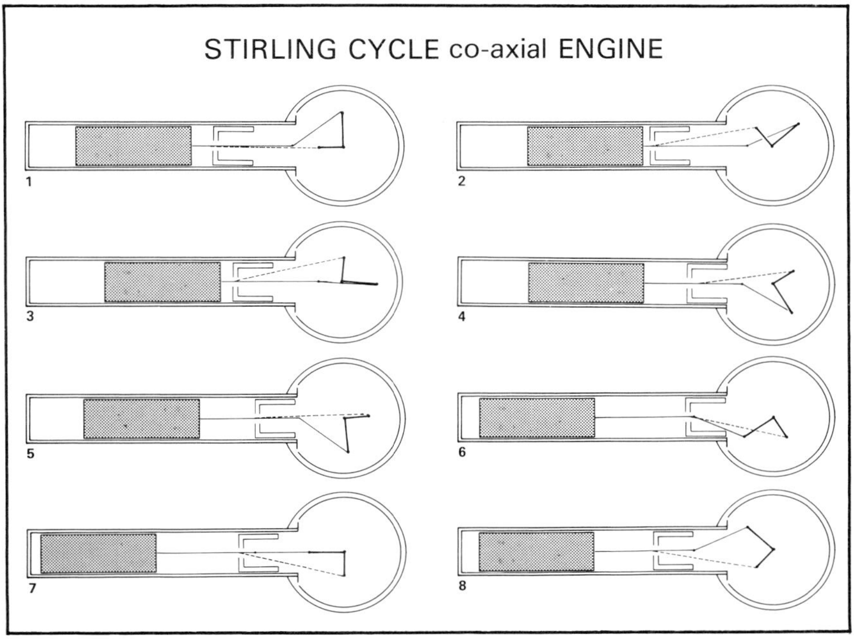

STIRLING CYCLE co-axial ENGINE

4 With the same scale draw two circles from the crankshaft centre, the inner circle representing the power crank stroke and the outer circle representing the displacer crank stroke.

5 Mark the position of the displacer at TDC, ie, fully pushed in. Measure the distance between the displacer front end and a point on the outer circle nearest the cylinder plate (stage 7 on illustration fig 6.20). That distance represents the displacer rod and con-rod combined length.

6 Follow stage 2 and draw the position of the power crank and the displacer crank at 10 past 11 on a clock dial, the short arm on the inner circle representing the power crank, the long arm on the outer circle representing the displacer crank. In this position the displacer and the power piston are at their closest. Draw in the power piston with $1/32$in gap from the displacer ($1/32$in on the actual scale, more on an enlarged scale). Draw a line to represent the con-rod from the power piston crown to the power crank-pin. This represents the power con-rod length (although adjustment to include the gudgeon block length will be required at a later stage of construction).

7 Fill in the other stages of the illustration to check that the various measurements are correct, eg, power piston length etc.

*The most important point to remember during the design stages is that dead space adversely affects the engine performance.*

Fig. 6.20 gives the more important stages in the movement of the pistons and the relationship of the power piston to the displacer during certain movements of the crank. Unlike the internal combustion engine, where each cycle of operation consists of four distinct phases: induction, compression, combustion (ignition or power) and exhaust, the hot air engine's phases are not so distinct and one phase leads into the other. In designing such an engine one must therefore look at all the various stages of the movements to obtain a complete picture of the mechanical action inside the cylinder. The construction of a co-axial engine follows the pattern set out in Chapter 13, 'Prova', Chapter 14, 'Sunspot' and Chapter 15, 'Dyna'.

## SOME PRACTICAL DETAILS OF ENGINE DESIGN

There is a relationship between the power cylinder bore diameter and the power piston stroke, equivalent to that found in internal combustion engines. Generally the relationship follows one of the following three parameters. In a square engine the power piston has an equal bore diameter and stroke (1in diameter × 1in stroke). In an over-square engine the stroke is much shorter than the diameter (½in stroke by 1in diameter). In an under-square engine the stroke is much longer than the diameter (1in stroke × ½in diameter).

In a small model hot air engine built for experiments or demonstrations, the square arrangement is the most suitable; side thrust on the piston wall is roughly equal to the load on the crank and bushes. For high revolution engines the over-square arrangement is more suitable, since there is a reduction of friction in the piston movement due to the shorter stroke. The disadvantage is a higher load on the crankpin and main bearing. For higher torque engines, the under-square engine is more effective, particularly if a long con-rod is used.

Where possible long con-rods should be used with power pistons. The longer the con-rod the lower the side thrust due to reduced angular displacement. The same applies with the con-rod fitting from the displacer rod to the displacer crank pin.

'Compression ratio' in hot air engines is something of a misnomer and is really the difference between the volume of air swept by the displacer piston and the volume of air swept by the power piston. Is is assumed that originally Stirling must have found that the ratio of 1.5:1 was the ideal one by trial and error, although there is a scientific reason for this. With regard to model engines this compression ratio is by no means to be taken as dogmatic. Different engine designs may require different ratios — indeed engines of the same configuration sometimes vary; while one engine may require a 2:1 ratio, another is quite happy with a 1.25:1. There are yet other types of engines, such as the Rider engines, which have a compression ratio nearer to unity, ie, 1:1; yet the power obtained from such models is amazing (Andy Ross' Rider engines are typical examples of this case).

Engines should be constructed in such a way as to quickly dismantle if and when the need arises. There is nothing more infuriating than to have to dismantle a whole engine for a minor fault or adjustment. This applies particularly to the co-axial engine where the displacer and the power piston are in the same cylinder.

The variable factors determining the success or otherwise of a Stirling engine are so numerous that the scope for invention and mechanical experiments has been by no means exhausted. Many an experimental or test-bed engine has been built in school laboratories and other workshops, with a view to testing some of these variables.

In the first place the test-bed engine frame should be easily dismantled, with all main parts quickly removable. This should be done in such a manner that any one part under study can be removed or changed without disturbing the rest. As an example, the construction of Prova II, described in Chapter 13, followed this pattern. Similarly the main or working cylinder, the burner, cooler and the flywheel should all be capable of removal with the least possible disturbance to other parts. If it is decided for example, to experiment on displacers or moveable regenerators, as Mr D. Urwick experimented on his regenerators (see Chapter 3), the test-bed engine should have a common drive mechanism, proven to be reliable and efficient, and all experiments undertaken by the simple removal of the main cylinder, either by unscrewing it or removing the bolts from the flange. Provision also has to be made for ease of changing the drive mechanism should the need arise. Designing a test-bed engine is a most interesting exercise, worthy of careful consideration by any enthusiast who wants to learn more about the thermodynamic principles of the Stirling-cycle.

CHAPTER 7

# WORKSHOP PRACTICE

Building the simple model engine does not call for sophisticated tools or special skills. Building a hot air engine can be fun even with a few basic workshop tools. Rarely, however, does the story end there. Many modellers have gone on to building more advanced engines, to researching on the principle, and to designing their own models. The author built his first working model from a school textbook with nothing more elaborate than an electric drill on a stand in the workshop. Progress in building more model engines brought about better equipment and tools. Every engine is a challange not only in its performance but in building it with what equipment and materials are readily available. Since cost and availability of materials may be limiting factors, ways and means of obtaining good working materials from alternative sources with least cost to the hobbyist are discussed in this chapter. The construction of 'Dyna', described in Chapter 15, is an example of how to design and construct an engine with a minimum of tools and purchased material.

## WORKSHOP FACILITIES

Basic tools are : Bench drill or drill press; set of HSS drills; set of taps and dies; Conecut or fly cutter (or similar adjustable hole boring tool); set of spanners — open-ended, ring and tubular; 3in to 4in bench vice; drill vice; reamers — hand; normal workshop hand tools — pliers, mole wrench, cutter, hacksaws, screwdrivers, steel rule, hammer etc, etc; miscellaneous files; small G-clamps; vernier calipers, scriber, centre punch.

Expenditure on a number of workshop small tools can be reduced with some planning. For example, the more common size screws, bolts and nuts are used in these models, mostly 6 BA and 4 BA although sometimes 8 BA or 2 BA are required. Similarly, silver steel rods (for connecting rods and crankshafts) of ⅛in and ³/₁₆in are more commonly used since they are easier sizes to handle and to find. It follows therefore that HSS drills, reamers, taps & dies, spanners to be used with the above sizes will be quite sufficient to begin with, before decisions are taken on how and when to expand workshop facilities. One way of cutting down on costs is shopping by post from the many excellent mail-order firms that specialize in surplus engineers' supplies, materials and redundant stocks.

At some stage a beginner to model engineering will think about setting up his own workshop, however small, however modest, and invariably the first acquisition will be the lathe; not surprising since the model engineer's workshop revolves round a lathe. A great deal of thought should go into the choice. Type, size and economics — all have their considerations and generally some bearing on the choice. The first question, which only the hobbyist can answer, is 'how far is my hobby likely to go'? Far too many beginners invest in expensive tools only to find that they really have little aptitude, or no time, or have had to move into smaller living premises. One should also consider whether by joining the local model engineering club, some of the heavier pieces of lathe work can be done in the club workshop and the minor work done on a small model engineer's lathe at home. Work on model Stirling engines mostly involves the use of the smaller lathe with only the occasional use of a larger chuck if cylinders exceed the 2in OD. The 1in OD cylinder gives ample scope for experimenting. In addition, the size of the engine remains relatively small while material for cylinders, displacers and pistons is more readily available.

A model engineer's lathe should meet all his immediate and long term requirements. In looking for one among the many kinds available on the market one should compare the different features, centre height, distance between centres, range of speeds, ruggedness of headstock and tailstock, screw cutting (with capability to change from imperial to metric threads) and the possibility of adding attachments for extra tooling, such as milling and drilling, as well as other optional extras such as change of tool holders, or a bored spindle. One should look for bargains available during reduced stock sales and national exhibitions. Many a good bargain can be had from advertisements in popular engineering periodicals where used lathes and other equipment in very good condition, made by reputable firms with household names are advertised for quick sale at greatly reduced prices. In the case of used equipment, a beginner should always seek the advice and assistance of the older members of the local club or others with engineering experience.

Except for the odd job in the workshop, there is not much call for the electric arc welding set, certainly not at the beginning. But the same cannot be said about oxyacetylene or oxygen-propane/butane-brazing sets. Although one can build the first few engines with the occasional help of a friend's kit, the need for this equipment gradually increases with workshop practice and usage. If the need is for the odd job around the house, some model engine work and such like jobs, then a small portable oxyacetylene or oxy-propane/butane set with rechargeable or exchangeable cylinders will be just the right type of equipment, available nowadays at very reasonable prices.

There are a number of engineering supply firms that specialise in selling workshop accessories in kit and castings which make valuable additions to the workshop equipment. Among the useful sets of castings one can find; bench and drill vices, power hacksaws, bench grinders, grinding jigs and angle plates. The advice here is to choose the more important and less complicated casting kit and work upwards with expertise.

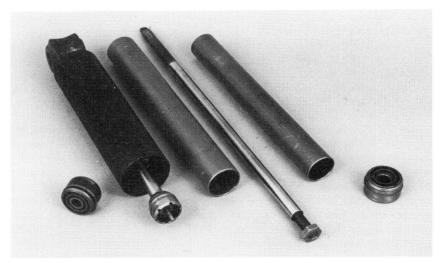

**Above** Fig 7.1 *Dismantled shock absorbers.*

**Left** Fig 7.2 *Scrap material used in a pressurised V-type Stirling engine.*

## SOURCES OF SUITABLE WORKSHOP MATERIALS

Shock absorbers (dampers) are a good source of material which can be used in the building of model engines (fig 7.1). Most shock absorbers have an internal cylinder or tube through which a plunger operates on hydraulic oil (and sometimes gas). A shock absorber that has not been grossly overworked or damaged can provide a number of materials that can be utilised. The internal cylinder can be used for displacer cylinders or even power cylinders. The metal, usually bright mild steel, is readily machinable giving an extremely fine finish. Most of these (internal) tubes from medium sized absorbers have an internal diameter of ⅞in to 1 ³/₁₆in which is ideal for most engines. Some of the larger shock absorbers have a heavy solid and single drawn outer cover which, with very little work, can be used for single cylinder co-axial engines of moderate size. Other shock absorbers have as plungers a piston with one or more compression rings. These pistons can be used with

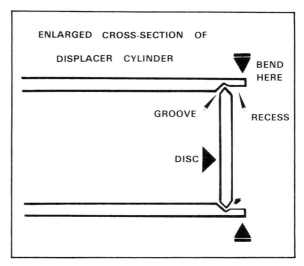

Fig 7.3 *Sealing a stainless steel cylinder without welding* (C. Vassallo).

ENLARGED CROSS-SECTION OF

DISPLACER CYLINDER

BEND HERE

GROOVE

RECESS

DISC

minor modifications, utilising the same rings or 'O' rings, as power pistons, provided that friction can be avoided. Other parts that can be salvaged from the smaller shock absorbers are piston rods, narrow-bore thin gauge pipes and chromed spindle rods.

A word of caution on opening shock absorbers; invariably these are sealed under pressure and should the absorber being dismantled still contain either gas or fluid under pressure, the sudden release can cause havoc to clothes, workbench and, very often, the ceiling. Always cut an opening with a very fine hacksaw blade some 1¼ in from the top end, keeping a cloth handy to cover the opening in the event of fluid escaping under pressure. Continue to cut around the circumference, taking care not to go too deep and damage the internal cylinder. Ensure that the plunger is completely extended out of the absorber so that any remaining fluid is contained at the bottom end of the absorber body.

Odd cuts of different diameter silencer pipes as well as the other casings of the larger shock absorbers make excellent ring burners (see Chapter 4).

Another source of material is from scrapped or surplus ex-services electrical equipment which can still be found or purchased for very little money. The amount of material that can be salvaged from some of this equipment is amazing, from high quality 2, 4, 6 BA bolts and nuts, to brass and dural rods and sections come many bits and pieces that can prove useful. Some component chassis make ideal bases for the model Stirling engine.

## THE MATERIAL COMPONENTS OF MODEL HOT AIR MACHINES

**Displacer cylinders or working cylinders** of single cylinder engines are best made out of stainless steel. When this is not available, bright mild steel tubes are a good close second. Ideally stainless steel cylinders should have their closed ends TIG-welded or MIG-welded and therefore capable of withstanding very high heat. TIG welding facilities are not readily available to all model engineers. The alternative is bright mild steel, brazed with iron fillings rod or arc-welded. This type of welding also withstands considerable heat. Brass brazing, although suitable for most small hot air engines, does not stand as much direct heat, especially from gas burners.

A novel way of closing a stainless steel cylinder end without any kind of welding has been evolved by Mr Cristinu Vassallo of Malta and used for all his Stirling engines (fig 7.4). This method involves delicate lathe work and bench fitting. The rear internal end of the stainless steel cylinder has a V-groove recessed, by machining in the lathe, to .045in ($^3$/₆₄in), .030in (1/32in) inside from the end. A stainless steel disc, only very slightly smaller in diameter than the total internal diameter of the recessed part (ie, the diameter of the cylinder + .045in + .045in), is prepared on a grinder so that the edge is ground all round the circumference into a V-shape. The disc is placed inside the V-groove, and the overlay very, very gently hammered by small taps of a light hammer into a folded seam over the disc. A salt solution is prepared, one tablespoon of kitchen salt just liquified with water, poured inside the cylinder and left standing overnight. On the following day the cylinder is cleaned internally and light oil applied until the cylinder is ready for use. The chemical action together with the fine machining gives a completely airtight cylinder capable of withstanding gas heat. (The finished cylinder end can be seen in the multi-lever twin power cylinder engine in Chapter 6).

Another novel method of closing displacer cylinder hot ends is used by Professor Charles Camilleri of Malta. This method also requires high precision work and involves the cutting of an internal fine thread (about 40 tpi) in the displacer cylinder hot end and an external thread of the same size in a ³/₁₆in brass disc which then fits snugly and exactly in the cylinder. The brass disc is additionally machined on the internal face to reduce thickness of material and on the external face to take a slot for a screwdriver. No bonding or other sealing is necessary. When screwed tight and heat is applied the expansion of the brass disc is sufficient to prevent any escape of the working gas.

After welding, brazing or otherwise closing the hot end, the cylinder is thinned down by machining to .015in (¹/₆₄in), certainly not exceeding .030in (¹/₃₂in), to cut down on heat conduction along the cylinder wall. Displacer lengths (except for single cylinder engines) may be broadly calculated at four to five times the diameter; single cylinder engine lengths depend on the layout and the drive mechanism employed.

**Power cylinders** can be made from different materials, stainless steel, bright mild steel, brass or even pyrex. The most important requirement is the fine internal finish, first through machining and then honing. The length of power cylinders depends on the engine layout, the piston length and the stroke. In single cylinder engines where the front end serves as the power cylinder, the internal finish should be just as fine as with any other power cylinder. Hand lapping is essential to give a mirror finish (see Chapter 8 for lapping).

**Displacer and power cylinders** are generally fitted to the engine frame in one of two ways. Flange fitting requires the brazing of a plate to the cylinder,

Fig 7.4 *Displacer cylinder with brazed flange and flange to fit displacer rod guide bush and air-vent connection.*

drilling the bore before or after fitting. The flange plate, usually of bright mild steel should be at least $^1/_{16}$in thick, preferably $^3/_{32}$in. Care should be taken during the heat application stage to avoid distortion of the cylinder. Thread fitting requires the cutting of a fine thread on two components, an external thread on the cylinder and an internal thread on the receiving engine frame or cylinder plate. In this part of the operation if the cylinder plate is of bright mild steel or of brass, another method can be adopted; a brass ring ($^1/_8$ in wall), is threaded and brazed (silver brazing is sufficient) to the engine frame or cylinder plate. If dural plate is used ($^3/_{16}$in to $^1/_4$in), this can be threaded internally by placing the plate in a four-jaw chuck.

The 'heat and freeze' method described below can be used quite successfully in the fitting of flanges and power piston 'closed-ends'. 'Sturdy', described in Chapter 11, has four flanges, two for displacer cylinders and two for power cylinders, fitted with this method, while the two power cylinder closed ends are similarly fitted on. This system is particularly useful for those modellers who have limited brazing facilities. At the same time the fitting of flanges by brazing or welding invariably causes cylinder warping and distortion by the application of high heat.

Basically the method involves freezing one component, heating the other and putting the two together with some bonding agent. Usually the metals used are different, dural and bright mild steel, for example. Dural has a high coefficient of expansion while mild steel has a low coefficient of expansion. The dural flange has a hole bored which does not quite take the mild steel cylinder. Cooling the cylinder in a freezer for one hour to $-20\,°$C and heating the flange by gas flame gives sufficient contraction and expansion difference to obtain a good, sliding fit. A smear of bonding agent to the lip of the cylinder before insertion ensures secure fitting against pressures. The dural plate should be ³/₁₆in to ¼in thick to prevent warping and provide a strong flange.

**Displacers** are light in weight and airtight, unless a displacer of the mesh kind described in Chapter 3 is used. Normally they are two and a half to three times the diameter in length. Thus a 1in OD displacer is about 2½in to 3in long. The fitting of displacers in the displacer cylinder is critical. Too narrow a gap between the cylinder wall and the displacer will cause a damping effect; too wide a gap will give too much dead space and no regenerative effect. The same applies if the gap at the top and bottom of the stroke of the displacer is too wide. Clearance around the displacer is normally of .015in (¹/₆₄in on radius) or .030in (¹/₃₂in on diameter), with the same gap at both ends.

In very small models, where it is not possible to make very light-in-weight displacers, the use of aluminium containers such as used for cosmetics or medicine, is an acceptable alternative. However the regenerative effect of aluminium displacers is limited. Once the displacer becomes hot throughout its length, the regenerative effect is lost, the difference in temperature between the hot and cold ends narrows and the engine efficiency falls drastically.

Where possible, displacers are made of bright mild steel, brazed at one end and plugged at the other by a lightweight plug and made airtight. The outer wall of the displacer is slowly thinned down by facing it with a number of fine cuts while in the lathe. The result is well worth the effort and time spent on this job. Displacers are checked for airtightness by placing in hot water and looking for tell-tale bubbles.

**Displacer con-rods** are normally made up in two parts, the part that fits into the displacer and the con-rod. In between there is usually a gudgeon block arrangement. The top part of the displacer rod is screwed tightly into the displacer plug end. In the course of running, it slides in the guide bush several hundred times a minute. This part is usually made from high quality silver steel, ground steel or polished (chromed) spindle steel, depending on the size of the model and the power output expected from it. An ordinary small model engine built for pleasure or minor experiments works well with a silver steel rod. On the other hand a competition engine would do better with a polished or chromed rod.

The con-rod attachment is made from bright mild steel or brass; at one end the con-rod fits into a gudgeon block fitted to the displacer rod, while at the other end the con-rod fits into a crankpin or similar arrangement. The con-rod can also be a combination of a small-end and a big-end with a threaded connecting rod between the two. This arrangement is quite handy for minor adjustments which may be required in the final fitting of an engine. The small-end of the con-rod fits into the gudgeon block while the big end usually has a bush or a ball bearing fitting to the crank. This type of con-rod is suitable for the larger type of model engine. In the case of small engines with a short stroke to the displacer, a Scotch yoke fitting can be made to the displacer rod.

**Guides or guide bushes** (fig 7.5) are made of brass or bright mild steel. The guide is made fairly long and is accurately machined in the lathe. It is centre-

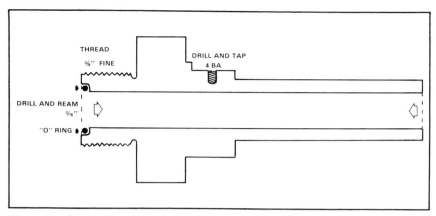

Fig 7.5 *Displacer rod guide bush.*

drilled in two stages (in the first stage by a drill slightly smaller than the final cut) and then finished off by a reamer. The length of the guide varies according to the size of the engine and the weight of the displacers. A long guide tends to create friction while a short guide has a double negative effect, possibility of easier escape of gas and faster wear on the displacer rod, particularly if the displacer is slightly heavy.

The displacer rod and guide assembly is a frequent area of failure and a reason why engines refuse to work or work sluggishly. Normally the two extremes are the cause. Too tight a fit creates friction or too loose a fit allows the escape of gas and consequent loss of pressure. The latter can be detected by the application of lubricating oil on the rod and listening for the hiss of escaping gas. Sometimes an engine which has refused to work, will do so on the application of a few drops of medium oil, but only until the sealing effect diminishes.

Guides of co-axial engines are machined somewhat differently although the drilling and reaming operation is common. These guides are either inserted in or are part of the power piston and are used at the same time as gudgeon blocks or yokes to the power con-rods.

**Gudgeon blocks and gudgeon pins** (known also as yoke) are the names given to a connecting point between two moving levers, one of which usually moves in a fixed direction, such as a piston or a displacer and rod, while the other is a lever that can turn through 360°. This arrangement can take different shapes, but for the purpose of this book, the fitting is made of a block. Whether a round bar or a square bar depends on the engine requirements. All the engines described in Part 2 have one or other type of gudgeon block fitting. Gudgeon blocks are machined such that the displacer rod is either screwed into or slides into one end of the block. In the latter case a set screw is used to retain the displacer rod. The other end of the gudgeon block has a slot cut into it for some length into which fits a con-rod without binding or loose action (fig 7.6). The retaining pin is called a gudgeon pin (or sometimes the king pin or wrist pin). Gudgeon blocks are machined from

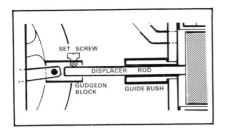 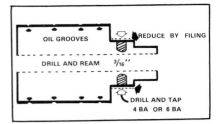

**Above left** Fig 7.6 *Sliding fit for a displacer rod into gudgeon block, tightened by a set screw.*

**Above right** Fig 7.7 *Power piston prepared with oil grooves as a compression aid.*

brass or bright mild steel, although any other material such as perspex, or PTFE may be used. The pin can be a bolt, preferably with a straight shank or a cotter (split) pin.

**Power pistons** (fig 7.7) are best made from cast iron or from an alloy which has a relatively low coefficient of expansion. Modellers have experimented with other materials such as PTFE and Rulon and have had highly successful results. One particular modeller, Mr F. M. Collins of Elstead, Surrey, has described and illustrated his method of machining PTFE power pistons *(Model Engineer*, June 20 1980, page 757) for his award winning engine 'Whippet' and also for 'Phoelix' described in the issue of August 3 1979. Brass is sometimes used in model engines and has been found quite suitable. Bright mild steel bars or pipes brazed at the closed end are also used. A piston made from old cast iron conduit water pipe, brazed and machined has given long and valuable service in one of the engines ('Dyna') described in this book.

The length of the power piston is a matter of calculation and depends on the diameter of the piston as well as on the stroke being used; generally speaking a length of ¾in to 1in for the smaller engine and 1in to 1¼in for the larger models is suitable. Power pistons are usually grooved with fine cuts to serve as oil grooves and an aid to compression. Some modellers prefer to give one of the grooves a deeper cut and to fill this with a PTFE twisted strand. Excellent compression can be obtained with this method. Yet another compression aid is the use of 'O' rings. The fitting of these is critical and requires good machining techniques and calculations. An 'O' ring can make a power piston stick and create a substantial amount of friction. If too loose there is no benefit accruing. Even in the best fitting 'O'-ringed piston there may be no bounce or reaction from a power piston when heat is applied to an engine — simple because of the 'sticking' effect.

The power piston usually employs a gudgeon block or yoke to take the con-rod which will connect the piston to the crank pin, disc or web. The construction of the gudgeon block is not very different from that described earlier. It is only the fitting to the power piston which differs. There are two principal methods for this fitting — either using a screw from the piston crown to the gudgeon block (fig 7.8), or reducing and threading the end of the gudgeon block to fit into the similarly tapped crown of the piston. Either

**Right** Fig 7.8 *Power piston with screw-in gudgeon block.*

**Below** Fig 7.9 *Power piston with twin-conrod fitting.*

**Bottom** Fig 7.10 *Collection of flywheels—that on the extreme right is made of ash wood.*

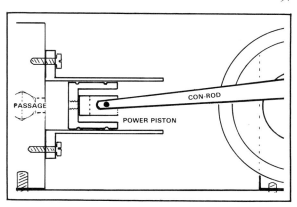

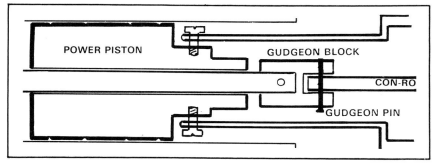

way requires that no gas will escape through this fitting. In co-axial engines the con-rod is assembled directly into or to the power piston (see displacer guide bush note above).

**Flywheels** (fig 7.10) for model hot air engines should not be heavy. Depending on the size of the engine a flywheel of two to three inches diameter, weighing between 8 to 12 oz is quite sufficient. Larger models, such as 'Dyna' described in Chapter 15 and the Ericsson engine described in Chapter 12, require a larger diameter flywheel, weighing about 14-16 oz. Hardwood flywheels, with a fairly thick rim, are an excellent alternative if metal ones are not readily available.

## BUILDING ENGINES FROM KITS OR TO PLANS

It is possible nowadays to buy the Stirling engine in kit-form and to complete it by machining and bench fitting at home or in the school laboratory. This method of building an engine has much to recommend it particularly when it is not possible or practicable to obtain the right materials or castings. The beginner to model engineering may have successfully completed one or two basic and simple engines and yet finds difficulty in going ahead with serious development because of such materials problems or the lack of proven advanced engine designs. This is where kit-form engines become useful alternatives.

One developer of engine kits is Mr Andrew (Andy) Ross of Columbus, Ohio, USA. Mr Ross, who in his professional life is an attorney-at-law and a great collector of historical Stirling engines and literature, is one of the foremost exponents and developers of the Stirling engine, with great machining expertise and successes particularly, but not solely, in the Stirling/Rider field. He has written several informative and very detailed articles, both in *Model Engineer* and in *Live Steam* (USA), and has published his own book, *Stirling-cycle engines,* which is well worth reading. Ross Experimental Inc of 1660 W. Henderson Road, Columbus, Ohio, 43220, USA, have produced more than one type of engine in kit form, the most popular being the B-20 (fig 7.12) incorporating the Ross linkage. Detailed plans for a V-type egine, the V-15, are available from the same address, also at a very reasonable price. Ross Experimental Inc also supply other products used in the building of model engines, such as stainless steel cups, steel shim etc. New engines are in the planning and development stage.

Fig 7.11 *Ross linkage constructed from a kit by Ross Experimental Inc.*

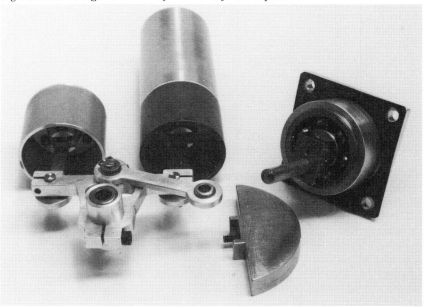

Fig 7.12 *Ross B-20 con-structed from a kit with a bicycle dynamo attach-ment.*

Other firms produce and supply engines in kit form or castings, advertising these engines in engineering periodicals in England, Europe and the United States. Solar Engines of Phoenix, Arizona, USA, advertise a selection of different Stirling-cycle and hot air engines, in kit form or as finished models, amongst which are a Heinrici-type engine, a Solar-engine and a Vacuum ('Flame Eater') engine. These models are of a very high standard and extremely well finished. Werner Wiggers of Kempen, West Germany adver-tise a beautifully designed 'Two-cylinder Hot Air Engine' which is available as drawings, castings or as a finished working model. Camden Miniature Steam Services advertise a 3in scale 'Rider-Ericsson' hot air pumping engine. This lovely and attractive model, authentic in design to the original engine, is available in castings with a construction manual. Edencombe Ltd of Middlesex, England, has just placed on the market an assembly kit for a Stirling cycle engine. The kit comes in the form of fully machinedand painted set of parts, requiring only screwdrivers and pliers for the engine assembly to be completed. A manual is included with the kit.

Building engines from home castings can be an alternative to kit construction. Basic engine frames and cylinder shells can be cast in a home or school workshop with relative ease and limited equipment. The advantage of such castings is principally one of compactness not possible with a bolt-on

**Left** Fig 7.13 *Home workshop aluminium casting of an engine frame.*

**Above** Fig 7.14 *Home cast engine frame for an experimental engine.*

**Below** Fig 7.15 *Stainless steel shim and mesh used extensively in regenerator experiments.*

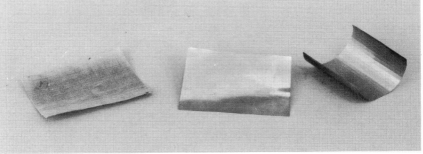

assembly (fig 7.13). Various shapes of engine frames can be designed and cast, while certain components are easier to cast and machine that to braze and face in the lathe. Such an item is the cylinder flange for both the displacer and power cylinders.

The use of steel shims as regenerative material is gradually gaining ground for two reasons. Steel is not a good heat conductor and heat deposited along a steel shim regenerator does not dissipate along the surface. It is readily picked up again by the working gas on its return. This is particularly useful in high speed engines. The second reason is that fine steel shim, .002in, occupies very little space while providing a large surface area particularly if the surface on both sides is used. Moreover if the regenerative gap is sufficiently wide and each layer is kept apart from the cylinder wall and from the next layer, more than one layer of shim can be inserted. This can be done by raising dimples (as with a seamstress tracing wheel—Andy Ross, *Model Engineer,* August 21 1981) or ruling it with a scriber along its length (as used in 'Prova II' in Chapter 13). Steel shim stock can be obtained in various thickness from .002in upwards. A few odd pieces of this material and some brass shim are always useful in the workshop.

CHAPTER 8

# STARTING AND RUNNING AN ENGINE

The author's very first engine started at the first attempt, but it cannot be said that all other engines were so cooperative. Indeed some of those built with great and loving care were quite frustrating until they started running, but once they did, they ran sweetly afterwards. Even the best designed and well constructed model hot air engines take some time to settle down.

Perhaps even more than an internal combustion engine or a steam engine, a hot air engine requires a long period of running-in, both during the various stages of assembly and in the initial period of its working life. Whereas the internal combustion engine has explosive power to counter the effect of friction during the running-in stages, a hot air engine has only the pressure of heated gas to make it turn. The running-in process involves four stages:

Stage 1: preparatory work prior to assembly.
Stage 2: running-in the assembled parts without compression.
Stage 3: running-in the engine under compression; but without heating.
Stage 4: running-in the engine under its own power.

## Stage 1: Pre-assembly

The power piston/power cylinder assembly is the part of the engine, irrespective of the type of engine, that is most liable to friction and that needs the greatest attention during the running-in process. No matter how fine a cut and how good a machine finish is given to the power piston and to the power cylinder, the final stages of finishing this particular assembly must be treated seriously.

**Lapping** may be done in the following manner: A wooden dowel of the same diameter as the piston, is dipped in paraffin until well soaked, smeared very lightly with extra fine valve grinding paste and inserted into the cylinder. The dowel is rotated slowly and evenly while at the same time moved in and out. Dipping in paraffin and smearing with the fine grinding paste continues until a silver mirror finish is achieved. The longer the job takes the better the finish to the bore. Apart from the silver mirror finish, perfect roundness is obtained by constantly turning the lap until there is no binding.

A wooden lap may be constructed in such a way as to allow for its expansion should there be excessive wear of the dowel. One way for this to be done

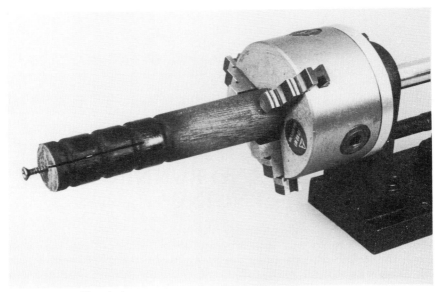

Fig 8.1 *A wood lap device.*

efficiently is to make a cut along the length of the dowel and to insert either a thin wooden wedge, or better still a long fine wood screw. The author has assembled a turning device for just this purpose which is extremely useful, practical, labour saving and efficient. The device (fig 8.1) has a detachable revolving chuck which is mounted to face the modeller. The wooden dowel, duly smeared, is held in the chuck which is made to revolve at slow speed, 100 rpm being sufficient for this purpose. The cylinder to be lapped is pushed in slowly along its length and then reversed until the lapping is completed. If the cylinder is open ended, any adjustment to the wood lap may be made with the cylinder completely in and the tip of the wood lap showing.

The power piston is best finished off while it is still in the lathe chuck. Once it has been ascertained that the fit is near perfect, the chuck is revolved at the slowest speed possible and extremely fine glass (or wet and dry) paper is applied along the piston length at one go. Only the slightest pressure is required to obtain a shiny and smooth finish. On ascertaining that the fit is perfect and after thoroughly cleaning all traces of grit with paraffin, Brasso or a similar polishing compound is applied with a soft cloth. Paraffin should also be liberally applied to the cylinder to ensure perfect cleanliness. The piston is run inside the length of the cylinder slowly turning it while pushing in and out. Any snagging or binding should be removed with polishing compound if the binding is slight, or by touching it with fine glass paper if the binding is extensive; this before any further assembly is attempted. Binding spots are easily discernible on the piston and in the cylinder wall. While this stage of finishing is in progress liberal use of paraffin is advised. When all binding is removed and piston action is sufficiently smooth, all traces of paraffin are removed and lubrication is achieved with a very fine oil.

Another part of the engine assembly that requires particular attention during the pre-assembly stage is the displacer rod and guide bush fitting. This is yet another problem area mainly because of the ease with which gas can escape through the assembly when the engine is under pressure. The displacer rod has always been a headache to modellers of hot air engines and although temporary remedial action can sometimes be taken by the inclusion or adaption of 'O' rings or lipped PTFE washers, the precision taken during the construction and running-in stages gives longer lasting efficiency than any corrective action later. The guide bush is drilled in one operation, drilling twice. First using a drill slightly smaller than the bore required, then using the right size of drill, and finishing off with a reamer while the bush is held in the chuck. When the guide bush has been reamed, the displacer rod is run in gently throughout the length of the guide several times by hand, having first smeared the rod with a very fine light oil.

Other moving parts of the engine, such as the crankshaft, connecting rods and crankpins should all be tried and turned by hand several times, and any binding, however slight, removed before the engine is assembled.

## Stage 2: The assembled engine

Once the engine has been assembled the second stage of running-in can begin. In order to facilitate the work and to avoid unnecessary strain on crankpins and con-rods, provision should be made for the working gas to escape at will. In parallel or separate twin cylinder engines this can be done by removing or loosening gaskets while in a single cylinder engine, such as the co-axial engine, an escape vent is drilled into the crown of the power piston and tapped to take a plug bolt.

During this second stage the engine is worked by hand — turning the flywheel to check again on any possible binding spot. Assuming all snags have been eliminated, the engine is turned several hundred times either by hand or by mechanical means. To this end three alternative methods are suggested: a larger diameter wheel (pram or small bicycle wheel with tyre removed) may be adapted into a pulley with an appropriate turning handle, coupled to a small pulley replacing the flywheel on the engine crankshaft and using plastic or rubber belting. The greater the difference between the size of pulleys the easier the work on one's arms. When the engine has bedded down slightly the revolutions can be increased. An alternative method is to couple a pulley on the engine crankshaft to an appropriate pulley on a lathe, keeping to low revolutions and taking care to avoid physical injury.

The method recommended by the author and which has been used on almost all the engines constructed involves a slow running electric mains synchronous geared motor that turns a mere 5 rpm (fig 8.2). Similar slow revolving motors that can turn between 5 to 30 rpm can be purchased very cheaply from surplus equipment stores. The electric motor is fitted with a pulley and the assembly bolted to the workbench. The model engine, with a small (2in) pulley replacing the flywheel, is mounted so that the two pulleys line up with each other, with a plastic/rubber belt between the pulleys. With such equipment the author has often left an engine running-in for several hours with an occasional drop of light lubricating oil. Obviously, other

Fig 8.2 *A slow turning running-in device.*

methods may complete this type of running-in, provided the desired efficiency is achieved.

## Stage 3: Running-in under compression

So far the engine has been turned without the benefit of internal compression and by now the moving parts have bedded. Any binding has been removed and the crank and con-rods are smooth in their movements. In this stage the compression is resumed by reversing the method used in stage 2, ie, replacing and tightening gaskets and plugging any air vent. The engine should immediately stiffen in its movements and back pressure should be felt as the flywheel is turned by hand. It may be possible to turn a small engine under compression by a mechanical means described in stage 2, but with a moderate size engine this may be difficult.

This stage of running-in requires only a relative short work-out if sufficient effort has been expended in the previous two stages. However patience is now required as the natural impulse is to skip any further time-consuming exercise and to start the engine. Experience has taught the author that this and the next stage not only ensure the successful completion of the running-in process but also a long trouble-free work-out. The few hundred revolutions required in this stage can be performed with the aid of simple turning device which can be used on almost all types of hot-air engines. This device consists of a lever, one end of which is slipped over the engine shaft and tightened with a set screw while the other end is used to turn the lever by means of a free revolving knob. This lever need not be longer than 4in to 6in between centres. Once this exercise has been performed, various parts are tested again for fitting and efficiency.

## Stage 4: Running-in the engine under its own power

Once heat is applied, the flywheel is flicked in the direction it has been designed to go. At first the flywheel will only complete one or two revolutions, but these revolutions will increase as the pressure of heated gas increases until sufficient energy is created to overcome the last remaining friction to turn over the flywheel. Most engines will complete a few revolutions and then stop. Experience has taught that if after one or two attempts to restart the engine there is no sign of life, the best thing to do is to let the engine cool completely and reheat after an hour or so. It will be found that in successive attempts the engine will work for longer periods until there is no further trouble in starting and getting the engine to run for a few minutes at a time. It is during this period that one's enthusiasm should be contained and the engine run for just a few minutes at a time between complete cooling. Indeed during the first few days a new engine should be run for two or three short periods an evening progressively increasing the duration of the runs as the engine settles down to smoother working and increased speed.

One word of warning; there is a tendency to increase heat in order to increase speed. This should be avoided in the initial stages. On the other hand, the cooling system should be at its most effective in these early runs. It appears that when an engine has reached its peak of efficiency, the amount of heating and cooling required to obtain the maximum speed and torque, level off to moderate amounts.

## ENGINE TUNING

An engine may appear to be giving its best performance yet the odd adjustment may bring about even better results. In engines where variable phase angles can be readily changed, tuning can be attempted by altering the phase angle first one way, then another, with certain safeguards. It is assumed that the engine has bedded down and that the running-in is almost completed. Also, the engine revolutions have been read over and over again by a mechanical revolution counter, or by stroboscope or tachometer. A clear indication that engine tuning is successful can only be obtained by reading off and comparing readings between one tune-up and another. The variations can be so slight that they will not be picked up by the naked eye.

The original position of the mechanism must be very clearly marked so that it can be returned to in the event that tuning is not successful. The phase angle is then increased, a few degrees at a time, certainly not more than 5°, preferably less. The engine is run a few times and readings are taken. If the readings are consistent, that is fine. If there is some variation, an average over the number of runs is taken. The phase angle is again reset, increasing by a few degrees, and readings are retaken. This should go on until either no further improvement is registered or the engine will not turn.

The original phase angle is returned to and further tests are taken, this time below the 90° phase by small changes in settings, again the smaller the better. The readings are noted down and comparisons made:

Original runs    600 rpm
   93° phase     610 rpm
   96° phase     615 rpm
  100° phase     620 rpm
  103° phase     580 rpm
  106° phase     no run
   87° phase     600 rpm
   84° phase     410 rpm
   80° phase     no run

Obviously the engine's efficiency in this case is best at the phase angle of 100°, giving 620 rpm.

With a twin-cylinder engine it is possible to change the compression ratio by altering one or both strokes of the power and the displacer con-rods, with appropriate changes to the length of the fittings and to the displacer body. These experiments are both useful and instructive. There are so many variables that influence the efficiency of the hot air engine that any experiments conducted on these lines will aid greater understanding of the thermo-dynamic principle on which the Stirling engine is based.

## MAINTENANCE AFTER RUNNING-IN AND TROUBLE-SHOOTING

As with all engines, once the running-in has been completed, a hot air engine should be cleaned of grime and used oil, the power piston wiped and very light lubricating oil applied to the piston, displacer rod and all working parts. All bolts and nuts should be checked and securely tightened and any gasket examined for the possibility of cracks or tears. Any special wear should be made up or parts replaced and generally the engine prepared as for demonstration purposes.

The accompanying charts have been prepared as an easy reference to the more common defects which can prevent the hot air engine from working or working efficiently. The illustrations (figs 8.3-8.6) are set out by particular problem areas so that the modeller can locate the defect or defects, cross-

Fig 8.3 *Friction diagram showing points of potential friction (1-8).*

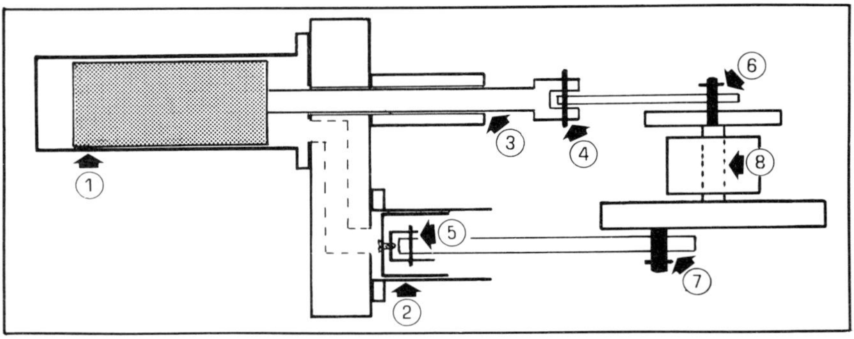

checking first against the general area which is suspect and then the particular trouble spot.

## Potential friction points:

1 Displacer scraping along the cylinder wall.
2 Power piston fit — too tight.
3 Displacer rod fit — too tight.
4 & 5 Gudgeon block/pin fitting of con-rods — too tight.
6 & 7 Crank-pin fitting of con-rods — too tight.
8 Crankshaft fitting in the bearing block — too tight.

**Friction** is the greatest 'killer' of hot air engines — any one of the points illustrated in fig 8.3, or worse still any combination of points, will prevent the engine from running. It is therefore important to ensure that all potential problem points are thoroughly checked during the early stages of running-in and before the final run. Friction points are the first suspect areas in any engine that will simply refuse to go, and not one of these can be disregarded.

Some defects leave tell-tale marks, others show binding during the free running. A scratching displacer can be heard, and if taken out of the cylinder can be checked for scratches along the surface. The power piston can be tested by a very simple method of blowing and sucking through the power cylinder end — the sliding fit should just permit the movement of the piston but not the escape of air. The displacer rod can be tested by inverting the engine and checking whether the weight of the displacer acting on the rod will allow an easy sliding movement without lubrication. The other points can be tested and checked visually for binding. The combined friction of different points on the cranks and con-rod fittings is enough to prevent an engine from running efficiently.

## Potential compression-loss points:

1 Displacer cylinder flange.
2 Power cylinder flange.
3 Power piston fitting.
4 Displacer rod fitting in guide bush.
5 Guide bush fitting in cylinder plate.
6 Power piston gudgeon block fitting.

Fig 8.4 *Loss of compression diagram.*

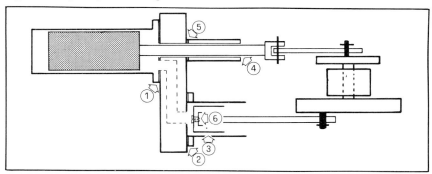

**Loss of compression** is the second major trouble area of Stirling engines. Hot air engines are low pressure machines and therefore the escape of a proportion of the working gas will prevent an engine from running. The major problem spots are shown in the illustration (fig 8.4). Some such as leaking flanges and cylinder plate fittings can be corrected quickly and without bother, while others may require additional machine work. A loose displacer rod fitting in the guide bush is best corrected by redrilling and reaming to the next higher sized silver steel rod available (eg, from ⅛ in to ⁵/₃₂in, while the gudgeon block fittings in a power piston can be corrected by the insertion of a fibre washer, bonding the surfaces and the retaining bolt with a smear of bonding agent.

The major problem point is the power piston fitting in its cylinder — a minor corrective operation is to machine a deep groove in the piston skirt or wall and to pack this with a PTFE tape twisted into a fine strand. Another alternative is the very precise fitting of a 1.5 mm 'O' ring. These are only corrective measures that can satisfy one's mind that the cause has been found. The real solution lies in building another piston with greater care and precision.

Compression losses are nearly always detectable by faint hissing noises at the various points when a very light oil is applied to the area. Better still, if the engine is placed in a shallow tray with sufficient paraffin to cover the engine, tell-tale bubbles will show the location of the trouble spots as soon as the flywheel is turned slowly by hand.

A quick and reliable test for the power piston is as follows. Place the cylinder, open end upwards on a piece of blotting paper with a few drops of water about the circumference. Place the power piston (with con-rod/gudgeon block attached), without any lubrication, just inside the cylinder. The behaviour of the piston is a good sign of its fitting — if it falls through the cylinder fairly rapidly, the fitting is too loose or there is a leak in the gudgeon block fitting and therefore a compression loss; if it does not budge even slightly with finger pressure, the fitting is too tight; if it sinks very, very slowly over a number of minutes — it has the right fitting.

## Dead space or dead volume:

1 & 2 Dead space at the top and bottom of the displacer stroke.
3 Dead space between the displacer and the cylinder wall.
4 Dead space in the connecting air passage.
5 Dead space at the top of the power stroke.

**Dead space or dead volume** may not prevent an engine from working but it will not work as efficiently as it should. If dead space is excessive the engine will probably work for a few seconds at a time, but never for long. Dead volume is completely unproductive, allowing heated gas to expand into the unoccupied space without giving a return for the energy expended on heating it. The major points of dead space are indicated in the diagram (fig 8.5) and are usually within the displacer cylinder where the greatest volume of gas is contained. The displacer fitting is critical and the clearance around it should be followed meticulously. For engines with a displacer cylinder of up to 1¼ in ID, a gap of ¹/₆₄in on the radius and ¹/₃₂in on the diameter is the ideal

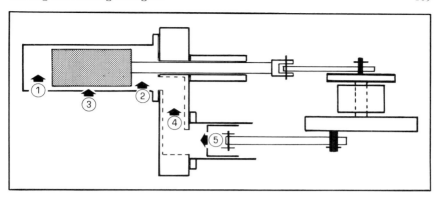

Fig 8.5 *Dead space diagram.*

clearance. A $1/64$in gap at the top and bottom of the stroke is also allowable if not desirable. Displacers made from aluminium containers expand on the application of heat and some may even form a dome at the closed (heated) end. This will result in a 'clacking' noise increasing in severity as the hot end becomes hotter. An engine may sieze if the increase in expansion exceeds the gap. This factor has to be taken into consideration. For engines with a light aluminium container as a displacer, the author generally allows slightly more than $1/64$in. If this is not sufficient, slightly thicker gasket material is used between the flange and the cylinder plate.

The air passage in twin-cylinder engines should be the shortest route between them, the diameter of the bore not exceeding ¼ in, preferably $3/16$in. The power piston at TDC should just about touch the closed end of the power cylinder; a tiny scrap of notepaper between the top of the power piston crown and the cylinder plate during setting will allow for any expansion of the piston when the engine has been working for some appreciable time at high heat.

Fig 8.6 *Diagram showing other problem areas.*

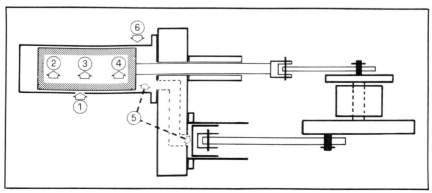

**Other problem areas:**
1 Too narrow a gap between the displacer and the cylinder wall — damping effect.
2 Displacer not airtight.
3 Displacer too heavy.
4 Displacer too short.
5 Compression ratio in need of revision.
6 Insufficient cooling.

**Other factors** will also affect the working of an engine. If the displacer fitting is too tight and the clearance is not sufficient for the working gas to push through with relative ease, the displacer will act as a damper. The fitting of a displacer can be roughly tested by the amount of air resistance it will meet during quick movements through the length of the cylinder. The displacer should be completely airtight — the slightest crack or smallest hole will prevent the engine from working due to the expansion of the working gas into the displacer — in effect it will be just dead space. A suspect displacer placed in hot water will leave tell-tale bubbles. A heavy displacer is a drag on engine power causing severe friction and sluggishness, while a short displacer, even with the right clearance in a short cylinder will not allow for the difference in temperature required between the hot end and the cold end to be maintained for anything but a very short period.

Another common failing point is an incorrectly calculated compression ratio. Although an engine may work on a ratio of 1:1 or 1:2, a smaller or bigger ratio will in all probability prevent an engine from working at all. Lastly insufficient cooling will stop the engine within a few minutes and should be considered as another failing point.

The most reliable engine test is the engine 'bounce'. When the flywheel is flicked slowly, it should not complete its revolution, but 'bounce' back more or less to its original position. If it is flicked sharply it will turn a couple of revolutions but again 'bounce' backwards and forwards a few times. An engine which has little compression or too much friction will not 'bounce'. Instead the flywheel may stop anywhere in its revolutions. Another indication of the efficiency of the power piston assembly becomes apparent whenever heat is applied to the engine. As soon as the engine 'feels' the heat there is an immediate reaction from the power piston which may move slightly backwards or forwards.

CHAPTER 9

# HOW TO CONSTRUCT 'DOLLY'

This engine (fig 9.1) is a basic model specially designed for beginners to the hobby. It is a compact engine and if well finished makes an ideal demonstration or desk-top model. The engine can easily attain a speed of 1000 rpm and will continue to run steadily as long as a flame is applied. A flame from a spirit burner or from a cube of solid fuel is sufficient. The construction allows for some minor but nevertheless interesting experiments. An up-scaled version of the engine can serve as a basis for research into the principles of thermodynamics. The design is pleasing to the eye, has clean lines and is uncomplicated in construction. It poses no problems to the modeller or to anyone interested in learning how the Stirling engine works. Only basic workshop tools are required but access to a small lathe and to brazing equipment will ensure trouble-free working and a good finish.

Fig 9.1 *Completed engine.*

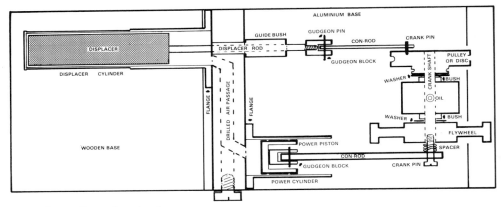

Fig 9.2 *General engine layout.*

## GENERAL ENGINE LAYOUT

The engine consists of two cylinders placed parallel to each other and bolted on opposite sides of a cylinder plate. One of the cylinders houses the displacer, while the other is the power cylinder. The displacer and power piston con-rods run parallel to a central drive mechanism consisting of a flywheel on the power side and a disc or web (or a pulley) on the displacer side. The expansion of gas and increased pressure are transmitted from one cylinder to the other through a 'hidden' interconnecting passage drilled through the cylinder plate in such a way that gas is well contained between the two cylinders. Heating is by a methylated spirit burner or solid fuel, while cooling is by the fin method. A larger scale model can be heated by gas and cooled by a water jacket. The design and construction allows for two parameters to be changed and experimented upon — the phase angle and the ratio of volumes.

## METHOD OF CONSTRUCTION

**The cylinder plate** is made from ⅝ in dural, polished, marked and drilled. On the power side the plate is drilled to take the displacer rod guide bush, the power cylinder flange and, at an angle shown in fig 9.2, the power side of the inter-connecting air passage. On the reverse, the displacer cylinder side, the plate is drilled to take the displacer cylinder flange and the inter-connecting air passage. The edge of the cylinder plate, on the power cylinder side, is drilled through for about three-quarters of the length, to meet the laterally drilled holes from the power and displacer cylinders (fig 9.2). This hole is tapped ⅜ in BSF to take a plug in the form of a bolt with a fibre washer to seal the passage. The displacer guide bush hole is drilled and tapped O BA, the holes for both flanges are drilled and tapped 4 BA while the inter-connecting passages are drilled ¼ in. The cylinder plate is also drilled and tapped 2 BA for bolting to the dural base.

   **The power cylinder** is open at both ends with a flange brazed or bonded on ('heat and freeze' method — see Chapter 7). The cylinder is made from thin-walled bright mild steel tube, honed and prepared to take the power piston.

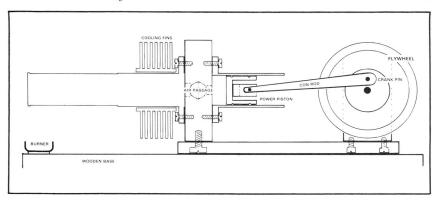

**Above** Fig 9.3 *Side eleva-tion—power side.*

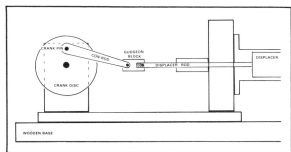

**Right** Fig 9.4 *Side eleva-tion—displacer side.*

**The power piston** is machined from a suitable piece of cast iron (brass or mild steel bar are good alternatives). Two oil grooves are cut into the piston wall. A precision fit is essential in the power piston/cylinder assembly; if this is obtained then the engine will almost certainly run without trouble. (A detailed method of lapping is explained in Chapter 8). The power piston should be a gas-tight fit, friction free yet sliding in and out easily without binding. Time consumed on this operation is never wasted. The gudgeon block is made from brass rod, slotted one end to take the con-rod and drilled and tapped 2 BA at the other end to take the securing screw which fastens the gudgeon block to the power piston. A $1/16$in hole is drilled laterally in the slotted part of the gudgeon block to take the gudgeon pin. The con-rod is shaped from $1/8$in bright mild steel strip, drilled at one end to take the gudgeon pin and at the other end to take the crank pin.

**The displacer cylinder** is made from bright mild steel tube, brazed or welded at the closed end and flanged at the other end (brazed, or 'heat and freeze' method). The displacer cylinder is further machined to reduce the cylinder wall to at least 50 per cent of its thickness, more if adequate tooling permits for this operation.

**The displacer** is made (or cut) from a felt pen aluminium container with an aluminium or dural plug drilled and tapped 6 BA. An allowance for expansion of the container should be made in marking out the length of the displacer. Calculate $1/32$in from the hot end and $1/64$in from the cold end, a total of $3/64$in less than the displacer stroke.

An alternative container may be used if this is available; a lightweight thin gauge mild steel cylinder is even more desirable in view of the heat conduction. The most important requirement is that the annular gap, that is the gap between the displacer and the cylinder wall is narrow and does not exceed $^1/_{64}$in in radius or $^1/_{32}$in diameter.

**The displacer rod guide bush** is machined from $^3/_8$in bright mild steel or brass rod, reduced to $^5/_{16}$in at one end for $^1/_2$in, which is then threaded O BA. The bush is centre-drilled, drilled and reamed $^1/_8$in to take the displacer rod. A single operation is advised for this assembly.

**The drive mechanism** consists of a central dural block, $^5/_8$in wide, 1in. long and 1$^3/_4$in high. A $^5/_{16}$in hole is drilled laterally to take two brass bushes which serve as bearings to the crankshaft. (An alternative method is to use brass tubing which gives a good sliding fit to the crankshaft and therefore the dural block is drilled according to the outside diameter of the brass bush/tubing).

The dural block is drilled from the top to the middle of the crankshaft bore to provide for an oil hole for lubrication purposes. The hole is drilled $^1/_{16}$in, the upper part of it being enlarged by a countersink to accommodate a few drops of oil. An oil cup is an excellent alternative. The block is marked, drilled and tapped 2 BA to allow for bolting the drive mechanism to the base.

The flywheel is machined from 2$^1/_2$in bright mild steel or brass bar, centre-drilled and drilled $^3/_{16}$in. The flywheel boss is drilled and tapped to take a 6 BA grub screw. In machining the flywheel the outside boss should have an outside diameter of at least 1in to allow for various crank pin adjustments in the event of further experimentation. Initially a hole is drilled and tapped 4 BA $^1/_4$in from the centre of the flywheel thus giving a stroke of $^1/_2$in to the power piston.

The other part of the drive mechanism is a disc (flywheel or pulley) to drive the displacer piston. The diameter of this disc can be varied between 1$^1/_2$in to 2$^1/_2$in. In this model a combined disc/pulley is machined from a 1$^1/_2$in dural bar. A V-groove is cut into the $^1/_4$in wide disc which has a $^3/_4$in × $^1/_4$in wide boss on the inside, (ie, facing the dural block). The boss is drilled and tapped 6 BA to take a grub screw. A hole is drilled and tapped 4 BA on the outside

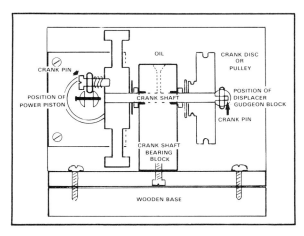

Fig 9.5 *Front elevation*

face of the disc, $^7/_{16}$in from the centre to give the displace a stroke of $^7/_8$in.

**An aluminium base** takes the cylinder plate and the drive mechanism block, the whole assembly being mounted on a wooden base which extends from the flywheel rim to the hot end of the displacer cylinder. Other metals, such as brass or zinc-based alloy, provide good substitutes. The metal base is marked, then drilled to take the bolts securing the cylinder plate and the flywheel block and also to take screws for bolting the base to the wooden base.

## STAGES OF ASSEMBLY

The cylinder plate is prepared with an inter-connecting air passage and plugged by a $^3/_8$in BSF bolt and fibre washer. The power cylinder is bolted in place ensuring that the air exit bore is completely clear of the flange or the cylinder wall. The displacer rod guide bush is finger tightened with a smear of Loctite. The displacer and displacer rod are inserted on the reverse side of the cylinder plate ensuring free movement of the displacer rod and perfect alignment of the displacer.

To ensure that the displacer cylinder is correctly positioned to avoid any internal friction between the displacer and the displacer cylinder wall, a piece of electrical tape may be wound round the top of the displacer (the hot end) and given two or three turns depending on the annular gap between the displacer and the cylinder. Since this gap should be minimal (see Chapter 6), it is not expected that more than two turns of the tape are necessary. Another piece of tape is similarly wound round the bottom end of the displacer (the cold end).

The cylinder plate is placed face down on a steady surface (or held tight in a 3in vice) and the displacer cylinder inserted over the displacer. Any adjustments to the tape thickness can be made before proceeding further. The cylinder, placed in this manner over the displacer, will ensure that there is no friction during the displacer movements. The position of the flange holes are marked on the cylinder plate, these are drilled and tapped 4 BA. The tape is removed, and the cylinder plate bolted on with a thin paper gasket between the flange and the cylinder plate. The position of the drive mechanism block is marked and the necessary holes are drilled for bolting the block to the dural base.

The drive mechanism is then assembled. The $^3/_{16}$in crankshaft is placed in the brass bush and the flywheel mounted with one or more washers to give the correct spacing. The disc/pulley is mounted on the other end of the crankshaft with spacers as required.

**Con-rod length calculations** The following method can be used to ensure correct lengths of the two con-rods. It is assumed that the con-rods have been drilled at one end only, the gudgeon block end, and that the other ends are unmarked and undrilled.

The flywheel is turned so that the crankpin hole is nearest to the power cylinder. The power piston is pushed in as far as it can go and the con-rod is marked at the centre of the crank-pin hole. An allowance should be made for any slight expansion that can take place to the power piston after this has been working for some time. A small piece of paper, one layer only, is inserted between the piston crown and the cylinder head. The paper is removed afterwards. The con-rod is drilled to take a 4 BA bolt and brass bush.

The same procedure is adopted for the displacer con-rod. The aluminium container displacer has a higher degree of expansion and therefore an allowance of $^1/_{32}$in should be made when pushing the displacer against the hot end of the cylinder and $^1/_{64}$in at the cold end. The con-rod is marked and drilled to take a 4 BA bolt and brass bush. The con-rods are finally inserted in the respective gudgeon blocks and the drive mechanism assembled.

**Timing the drive mechanism** The displacer is always 90° (a quarter of a turn) in advance of the power piston. Therefore if the engine is designed for the flywheel to turn clockwise, the setting of the drive mechanism should be done in this manner:

The flywheel, connected to the power piston, is turned so that the crank-pin is at its highest position from the base (fig 9.6a). That is the piston is half-

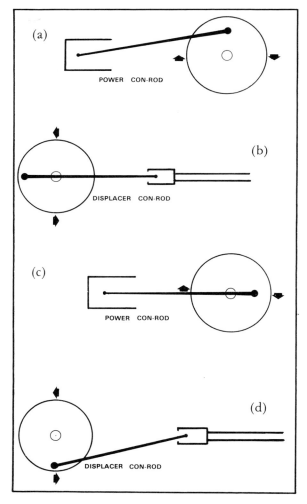

Fig 9.6 *(a) Power Piston. (b) Displacer. (c) Power Piston. (d) Displacer.*

way out from top dead centre (TDC). With the power piston in this position, the displacer disc is turned so that the crank-pin is furthest away from the displacer cylinder (fig 9.6b), that is the displacer is at bottom dead centre (BDC). In this position the displacer is within the cold space and also the greater volume of air (gas) is in the hot space. If the flywheel is turned a quarter of a turn clockwise, the power piston is now at BDC (fig 9.6c) while the displacer crank-pin is in its lowest position and nearest the base (fig 9.6d). This also means that the displacer is halfway into the hot space. The set screws are now tightened and the timing is set. If it is desired that the flywheel turns anticlockwise all that is required is for the displacer to be at TDC, that is fully pushed in, while the power piston crank-pin is at its highest position above the base, as in fig 9.6d. It may happen that an engine works better if the phase angle is slightly more or less than 90°. This is a matter for experimentation, see Chapter 8 under 'tuning'.

## PERFORMANCE

The engine ran after about 15 minutes of flicking the flywheel. However the engine bounce and the flywheel response indicated from the beginning that there was sufficient 'feel' in the engine to show promise. The first run lasted about two minutes but the engine would not run again until it had cooled down completely. Eventually the running time increased as did the revolutions. After a modification to the displacer cylinder by thinning down the hot end considerably, the engine response to heat became faster and the revolutions picked up quite substantially. Speeds in the region of 1000 rpm can be obtained from this engine provided that the power piston assembly is a gas-tight fit; the displacer guide bush/displacer rod assembly has a good sliding fit; there is no friction in the bearings, crankshaft, con-rods assembly or any other moving parts and that there is a free flow of gas between the power cylinder and the displacer cylinder.

The engine can be scaled up to about twice the original size. Heating by gas and cooling by water jacket will obtain a good performance from such a model, especially if a thin-walled light mild steel displacer is used. In scaling up the size of the engine, careful note should be taken that the displacer length/diameter ratio of 3:1 is, as far as possible, maintained. Whereas in smaller models a ratio of 4:1 is acceptable and sometimes desirable, an enlarged version displacer decreases in efficiency if the ratio of 3:1 is substantially altered.

**Engine specifications**

| | |
|---|---|
| Aluminium base | 5in × 3½in × ¼in |
| Dural cylinder plate | 3½in × 2¼in × ⅝in |
| Displacer cylinder | 3 $^{9}/_{16}$in external length |
| | 3½in internal length |
| | ¾in OD |
| | $^{21}/_{32}$in ID |
| Displacer cylinder flange | 1½in × 1⅝in |
| Displacer | ⅝in OD |
| | 2 $^{19}/_{32}$in length |

| | |
|---|---|
| Displacer rod | ⅛in silver steel |
| | 2⅞in to gudgeon pin centre |
| Displacer con-rod | 1 $^9/_{16}$in between centres |
| Power cylinder | 1½in length |
| | ¾in OD |
| | $^{11}/_{16}$in ID |
| Power cylinder flange | 1⅛in × 1½in |
| Power piston | ⅞in length |
| | $^{11}/_{16}$in OD |
| Power piston con-rod | 2⅝in between centres |
| Guide bush | 1⅜in length |
| | ⅜in OD |
| Flywheel | 2⅜ OD |
| | ⅜in rim thickness |
| Displacer drive disc | 1½in OD |
| | ⅜in thickness |
| | 1⅜in pulley diameter |
| Central dural block | ⅝in width |
| | 1in length |
| | 1¾in height |
| Cooling fins (five in number) | 2in OD |
| | ¾in length |
| Power piston stroke | ½in |
| Displacer stroke | ⅞in |

CHAPTER 10

# HOW TO CONSTRUCT 'LOLLY', A V-TYPE ENGINE

This engine (fig 10.1) is another model designed for beginners. The construction method is simple and the components are relatively easy to find or construct. It can be built with basic workshop tools while requiring limited brazing facilities.

The principle of the V-type Stirling engine embodies a single crank and crank-pin thus reducing the number of moving parts and consequently, friction. In a V-engine the power and displacer cylinders are at 90° to each

Fig 10.1 *Completed engine.*

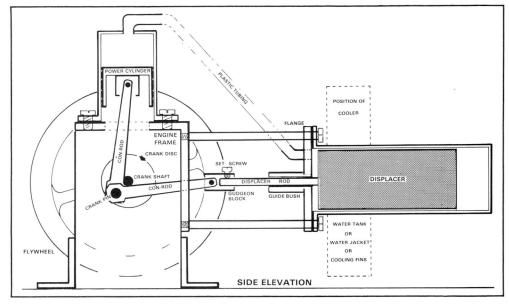

Fig 10.2 *Engine layout.*

other; this format can take different positions in relation to the engine base or frame. The V-type engine is one of the more efficient model engines for beginners since by its layout two parameters are automatically absorbed in the design and construction: the 90° phase and the ease of the working gas flow between the two cylinders. The position of the cylinders allows a variety of construction methods and the use of different metals and materials for the engine frame. One such engine seen by the author had the engine frame and cylinder supports made out of polished hardwood, which gave the engine a very elegant appearance.

## ENGINE LAYOUT

The engine has the power cylinder and the displacer cylinder placed at 90° to each other, the displacer cylinder is in a horizontal position with the power piston in a vertical position. Brazing facilities are required for the displacer, the displacer cylinder and the power cylinder. Additionally, though not essentially, the engine frame can be brazed to form a three sided box (half a cube), two adjacent sides supporting the cylinders while the third adjoining side supports the crankshaft bearing. The two cylinders are connected by means of copper and plastic tubing from the crown of the power cylinder to the base or flange of the displacer cylinder. Copper bends, ¼ in ID, can be soldered or bonded in place and a short length of plastic tubing inserted tightly over both ends giving the shortest air passage possible. The displacer cylinder and flange are bolted on four ¼ in pillars, as distant pieces, which in turn are screwed into the engine frame. This allows a long connecting rod between the displacer gudgeon block and the crank pin.

## METHOD OF CONSTRUCTION

**The engine frame** is prepared first. Mild steel flat bars, ⅛ in, or dural, ³/₁₆in, may be used, keeping in mind how the frame is to be mounted on to the base. The two cylinder plates are marked, drilled and the holes widened by a Conecut or similar boring tool. The cylinder plate which takes the displacer has the hole drilled ¼ in off-centre (see fig 10.3) to allow for the con-rod to fit the same crank-pin. The cylinder plate which supports the power piston is similarly marked, drilled and widened, allowing for the width of the crank-disc. The third side, adjacent to the two cylinder plates, is prepared to take a guide which will house the crankshaft. A long guide is recommended to support the flywheel and crank-disc and to avoid undue wear of the crankshaft or brass bearing. The guide is either brazed or threaded and bonded in place. The crankshaft bearing plate is brazed or bolted to the two adjacent sides to form half a cube. The engine frame is bolted to brackets shaped or cut from L-iron bars, which in turn are bolted to a wooden base. The displacer cylinder plate is marked, drilled and tapped with four 4 BA holes which will take the four pillars supporting the displacer flange and displacer cylinder assembly. These pillars may be machined from ¼ in hexagonal rods or from ¼ in silver steel rods, with the ends reduced and threaded 4 BA.

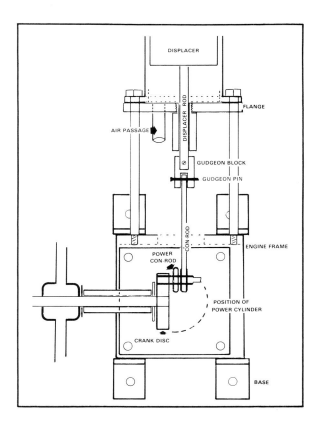

Fig 10.3 *Over view.*

**The power cylinder** is made from high quality bright mild steel. The closed end and the flange are silver brazed to the cylinder, which is then honed and lapped to a fine finish. (The 'heat and freeze' method may also be used with this cylinder. With this method there is less chance of metal distortion than with brazing). The cylinder head is drilled and fitted with a copper bend which has an internal diameter of ¼ in. (This type of bend can be purchased from ironmongers or from refrigerator repair shops that stock small bore copper tubing; these fittings are very useful in the construction of model engines and external regenerators). The bend may be sweat soldered, or silver brazed or even bonded in place.

**The power power piston** is machined from cast iron, or bright mild steel or brass. A tight fitting and a fine finish is essential for a successful engine. Two or three oil grooves are cut into the piston wall before the piston is removed from the chuck. A hole is drilled and counter-sunk to take the bolt which retains the gudgeon block. An alternative method is to drill and tap a 2 BA hole, into which the gudgeon block can be screwed.

**The gudgeon block** is machined from ½ in bright mild steel, drilled and tapped 2 BA at one end, and slotted ⁵/₃₂in at the other end to take the con-rod made from ⅛ in bright mild steel flat bar. The alternative method of screwing in the gudgeon block requires the round bar used for this block to be reduced to ³/₁₆in and threaded 2 BA, while the other end is slotted as before. The gudgeon block is drilled ⅛ in to take the gudgeon pin which will retain the con-rod. The other end of the con-rod is left clear for marking and drilling in the final fitting.

**The displacer cylinder** is best made from thin-gauge steel tubing. The alternative is bright mild steel, brazed (or welded) at the closed end, and the wall thickness reduced at the hot end. At the other end of the cylinder a flange is brazed and drilled with four holes which will take the pillars from the engine frame.

**The displacer** should be made from thin-gauge mild steel, brazed at one end and plugged by an aluminium plug, drilled and tapped 4 BA, at the other end. If such a size of cylinder is not available or cannot be machined from a suitable tube, an alternative source is an aluminium medicine or perfume container of the right size and diameter. A plug is prepared and bonded in place after being drilled and tapped 2 BA to take the displacer rod.

**Displacer flange and guide bush** This engine requires a second flange to contain and seal the displacer cylinder assembly. This flange actually replaces the cylinder plate which is normally an integral part of the engine frame. This flange is made sufficiently strong to withstand any warping that may occur when mounting on the engine frame pillars. Bright mild steel, ⅛ in thick, or dural ³/₁₆ in may be used.

Work on this flange includes the following: **1** marking and drilling the four holes which will take the four pillars from the engine frame. These should be aligned with the corresponding holes on the displacer cylinder flange; **2** marking, drilling and tapping the bore for the displacer rod guide; **3** marking and drilling the air exit bore which will connect the air passage from the displacer cylinder to the power cylinder — a copper bend as used on the power cylinder is to be inserted here.

Fig 10.4 *Side elevation.*

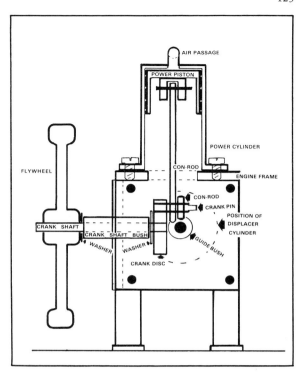

Particular care is taken when placing the displacer into the displacer cylinder and both flanges tightened together. The slightest mistake or miscalculation can cause friction between the displacer and the displacer cylinder wall. The guide and the copper bend are bonded in place and the assembly tightened with a gasket between the two flanges.

**A gudgeon block** is machined from ½ in bright mild steel, drilled and tapped 4 BA at one end to take the displacer rod, and slotted ⁵/₃₂in to take the displacer con-rod. A hole is drilled and tapped 4 BA to take the set screw which retains the displacer rod. This arrangement provides for fine adjustments to the displacer stroke. The slotted side is drilled ⅛ in to take the gudgeon pin.

**The con-rod** is cut and shaped from ⅛ in bright mild steel flat bar, drilled one end to take the gudgeon pin and prepared for marking and drilling at the other end in the final fitting.

**Crankshaft crank disc and crank pin** This size and type of engine should take a ¼ in crankshaft, reduced at one end and threaded 2 BA to take the crank disc. A suitable brass bush is inserted into the crankshaft guide to provide a good bearing surface. The crank disc is cut and machined from 1½ in bright mild steel bar, centre drilled, drilled and tapped 2 BA to take the crankshaft, and marked, drilled and tapped 4 BA ½ in from the centre to take the crank-pin. This will give the power piston and the displacer a stroke of 1in. The crankpin is machined from ³/₁₆in silver steel, reduced and threaded 4 BA both ends. One end is screwed into the crank-disc while the other end takes a 4 BA nut to secure both con-rods in place while the engine is running. The crankshaft and crank-disc are assembled and securely tightened with a smear of Loctite Screwlock 222. The crank-pin is only finger tightened at this stage.

## ASSEMBLING THE ENGINE

The engine frame is screwed or bolted to a wooden base, heavy enough to withstand a fair amount of vibration. Firstly the four pillars supporting the displacer cylinder unit are screwed into the engine frame, and well tightened. The bottom flange is placed over the four pillar ends; the displacer rod guide and copper bend are already bonded in place. A gasket, $1/32$in cork or similar jointing material is placed on the flange with suitable holes cut for the various openings. The displacer and displacer rod are inserted and the displacer cylinder fitted on. The four pillars are tightened with nuts and washers, care being taken that equal pressure is exerted on the four corners of the assembly. The free movement of the displacer is checked with each turn of the nuts to ensure correct displacer alignment inside the cylinder. The gudgeon block is then fitted on at the end of the displacer rod.

Secondly, the crankshaft assembly is inserted into the crankshaft guide, with washers where necessary, and the flywheel is fitted. The power cylinder, with the power piston and con-rod in position, is bolted on to the power cylinder plate. The power piston is pushed in as far as it can go (TDC); the crank disc is turned until the crank-pin is at its highest point, that is nearest to the power cylinder. The con-rod is marked where the crank-pin should fit, and drilled to take the pin with a brass bush as a bearing. The assembly is tested for stroke and compression.

The displacer (and rod) is pushed in as far as it can go, the rod is marked (by felt pen) where it stops at the guide bush; the displacer is then withdrawn $1/64$in to allow for expansion of the displacer ($1/32$in if the displacer is made of aluminium tubing — due to a higher degree of expansion of this metal). The crank-pin is then turned to the point nearest the displacer cylinder. The displacer con-rod is marked and drilled where the crank-pin should fit. The assembly is tested for movement and friction.

The next step is to insert the necessary washers and spacers between the crank-disc, con-rods and securing nut, ensuring that there is no binding in this assembly. The crank-pin should be well tightened in the crank-disc with a smear of Loctite Screwlock 222. A substantial running-in is undertaken at this stage making sure that there is excellent compression from the power piston assembly and there is absolutely no binding in any of the moving parts. The power piston can be tested for 'bounce' at this stage by placing a finger on the air exit hole.

The cooling assembly is fitted to the displacer cylinder. Cooling fins with a large surface area are adequate but only just, and only for short runs. A better type of cooling is called for if the engine is to be run for long periods. Cooling by water tank or by water jacket connected to a water tower can be recommended. Rubber or plastic tubing is inserted over the two air exit bends of the power and displacer cylinders. The engine is now ready for testing.

## RUNNING THE ENGINE AND MODIFICATIONS

Seen from the flywheel side, the engine turns in a clockwise direction; seen as in fig 10.2 the engine turns in an anti-clockwise direction — in each instance the flywheel is turning away from the displacer. If the fitting has been done carefully, there should be an immediate reaction from the power piston when the flame is applied to the displacer hot end, and the flywheel should move slightly in one or other direction. An engine of this size should have adequate heating preferably by gas burner, although a good sized spirit flame is sufficient to turn the engine, particularly when the engine has been running for some time and has settled down. Engines of this type seem to last longer as demonstration models if care and attention is given during the running-in period.

A number of modifications can be made if desired. The engine stroke can be modified by changing the position of the crank-pin, bringing this nearer the centre of the crank disc. For example a ⅜ in distance between centres will give a stroke of ¾ in. A short stroke will give this engine faster revolutions. Obviously the displacer is made longer to avoid dead space in the displacer cylinder during each stroke. Similarly adjustments are made to the power piston and displacer con-rods.

A minor modification can be made by changing the position of the cylinders, either by having the displacer cylinder in a vertical position (power piston horizontal) or by positioning the cylinder in a proper 'V' formation, that is, both cylinders 'in the air'. This latter position of the cylinders will probably result in higher engine revolutions as there is better distribution of the drive mechanism weight during the actual running.

This engine can be scaled up or down provided the ratios are maintained. A

Fig 10.5 *Alternative construction of engine frame.*

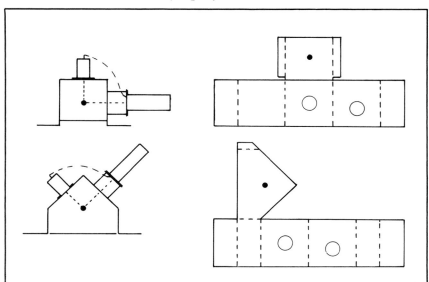

larger scale model can be quite powerful and made to perform work, such as pumping water to and from a tank. A small scale model makes an ideal desktop demonstration engine.

The engine frame can be made from one piece of sheet metal, preferably 1/16in bright mild steel, brazed, bolted, or even riveted. The sheet metal is designed, marked, drilled and bored as necessary, ensuring the correct positions of the cylinders and flange holes. The metal is cut as required and then bent over a vice or a suitable hard metal surface and gently hammered at the corners (fig 10.5). This method is suitable for engines with one cylinder in a horizontal position or with both cylinders 'in the air'.

The V-type engine has a fixed crank stroke for both cylinders, therefore the 90° phase is set automatically. There is room, however, for experimentation and for modifications during the construction stage for the angle between the cylinder plates to be altered to below or above the 90°. Another experiment that can be undertaken is the alteration of the volume ratios between the power and displacer cylinders to below or above the 1:1.5 ratio normally used. This can be done by altering the size of the displacer and the displacer cylinder. With careful planning much of the frame work and the drive mechanism, con-rods etc, can remain unaltered. This engine can take a miniature dynamo to light a small bulb. Such a dynamo can be wound and constructed from instructions found in simple school project books. Alternatively a small electric motor may be tried as a dynamo; one such motor is used in 'Dyna', Chapter 15.

## PERFORMANCE

This engine worked at first attempt and has been one of the most trouble-free models attempted by the author. After months of shelf life it still starts and goes without any bother as soon as heat is applied. This engine should easily reach a speed of 900 rpm once the parts have bedded down and any snags smoothed out. Occasional lubrication is required to the working parts and oiling of the surfaces to prevent any rusting. The gaskets should occasionally be checked for any cracks, while the pillar bolts may require a slight turn after the engine has been running for some time.

**Engine specifications**

| | |
|---|---|
| Engine frame | 3¼in × 3¼in × 3¼in |
| Engine frame legs extension | 4 of 1½in × ¾in |
| Power cylinder | 1⅝in OD |
| | 1½in ID |
| | 2½in length |
| Power cylinder flange | 2¾in × 2¾in |
| Power piston | 1½in OD |
| | 1¼in length |
| Power con-rod | 3 3/16in between centres |
| Displacer cylinder | 2in OD |
| | 1⅞in ID |
| | 5in length |
| Displacer cylinder flange | 3in × 3in |

| | |
|---|---|
| Displacer | 1 $^{13}/_{16}$in OD |
| | 3 $^{15}/_{16}$in length |
| Displacer rod | 2¾in |
| Displacer con-rod | 2¾in between centres |
| Displacer pillars | 4in × 3¼in (between threads) |
| Displacer flange (2nd) | 3in × 3in × $^3/_{16}$in dural |
| Flywheel | 5½in × ¾in rim |
| Crank-disc | 1½in OD |
| | ¼in thick |
| Crank-pin | $^3/_{16}$in silver steel |
| Crank-shaft | ¼in silver steel |
| Cooler | Water tank, water jacket or cooling fins |
| Burner | Gas |

CHAPTER 11

# HOW TO CONSTRUCT 'STURDY', A TWIN-STIRLING ENGINE

This engine (fig 11.1) is a compact twin Stirling engine designed as motive power for marine use or for land vehicles such as a truck. Originally the engine was designed to fit into a model cargo ship. This experiment was successful up to the point that the power generated was sufficient to move the ship at a steady and sedate speed. The problem with using this particular design in a ship was that the displacers were fairly high up and the burner, 7in above the keel, was subject to wind influence. On the other hand, the low centre of gravity makes it an ideal motive power for any vehicle which encloses the

**Left** Fig 11.1 *Completed engine.*

**Above right** Fig 11.2 *Engine layout.*

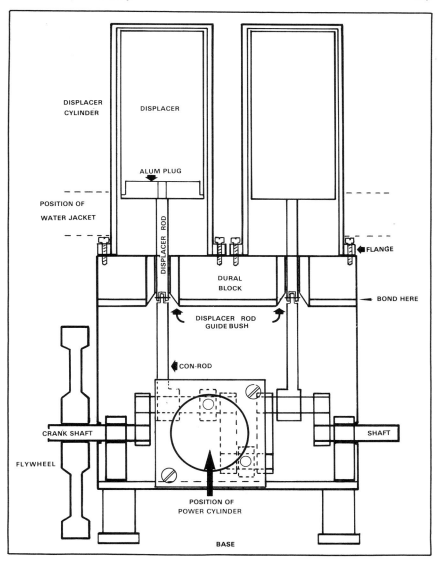

engine completely. The engine is constructed mainly from scrap material readily available, with little machining. It is well within the capabilities of a modeller who has successfully constructed the previous two models.

## GENERAL ENGINE LAYOUT

'Sturdy' is a combination of two separate Stirling engines built round one engine frame or compartment and having a common multiple crankshaft and flywheel. The engines are of the V-type with the displacers in a vertical position, one behind the other, while the power cylinders are in a horizontal position, on either side of the engine body. Each power cylinder is connected externally and directly to its displacer cylinder.

The engine compartment is constructed from an aluminium box obtained from scrap aluminium sections. The power and displacer cylinders are cut from two different sizes of internal shock absorber tubes while the displacers are taken from aluminium containers. Great care is taken in the construction of the crankshaft and in positioning the displacer and power con-rods.

As an experiment the 'heat and freeze' method of bonding flanges and cylinder ends is used extensively, replacing six brazing areas and using brazing only on the displacer cylinder hot ends. The flanges and closed ends are completely airtight to the extent that an attempt to pressurise the engine had positive results.

## METHOD OF CONSTRUCTION

**The engine compartment** is cut from scrap aluminium section (fig 11.3) normally used in the construction of aluminium doors and windows. These sections have a number of projections internally and externally which are removed by filing, leaving a neat box with internal dimensions of 2½in height, 1⅝in width and 3½in length. The compartment is marked and drilled in four places; two holes for the displacer rods and guides, and two holes for the power con-rods. The power side holes are further bored and widened by a Conecut to a diameter of 1in to allow for con-rod movement.

**The cylinder head or plate** is cut from ⅛in dural flat bar (fig 11.4) with the dimensions of the engine compartment top, 1¾in width and 3½in length. The cylinder head is marked and drilled ½in where the displacer rod guides are to be inserted. Accurate drilling is essential in this operation. The air-vent holes are drilled from above to half way through the thickness of the plate and then from the edges to provide for exit holes. Further work on the cylinder head takes place later when the displacer cylinders are fitted.

**The displacer cylinders** are cut from a shock absorber internal tube. The hot

Fig 11.3 *Extruded aluminium section.*

Fig 11.4 *Top view of dural block.*

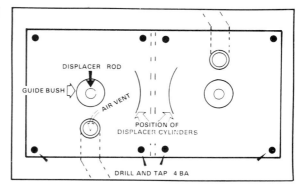

ends are brazed while the flanges are fitted with the 'heat and freeze' method. The normal method of brazing cylinder flanges may of course be used. The flanges are prepared with four holes for bolting to the cylinder head.

**The displacers** are cut to size from 'Albusticks' aluminium containers, and prepared to take aluminium plugs. The plugs are machined in the lathe, centre-drilled, drilled and tapped 4 BA to take the displacer rods. The plugs are bonded in place and the displacer rods screwed in with a smear of bonding agent. The displacer/displacer rod assemblies are checked for accuracy and perfect alignment.

**The displacer rod guides** are machined from ½in bright mild steel or brass rod drilled and reamed ³/₁₆in. Before the final reaming operation, the bottom end of the bore is widened by a countersink to accommodate the con-rod/displacer rod king-pin fitting. (This operation may not be necessary if a modification suggested by the author is followed — see end of chapter). The guides are inserted in the cylinder head and bonded in place leaving a projection of ¹/₁₆in to fit into the engine compartment.

At this stage the cylinder head is bonded to the engine compartment with a generous quantity of bonding agent round the projecting rod guides. This ensures that when the bond sets the cylinder head and engine compartment become one solid body. The next step is to fit the displacer cylinders to the cylinder head. In order to obtain perfect alignment and to avoid friction during the displacer movement, the method used in Chapter 9 (using tape round the displacer during the marking session) should be followed. The bolt holes are marked, drilled and tapped 4 BA, the tape removed and the displacer cylinders (with the displacers in place) assembled, and the bolts lightly tightened. The engine compartment is turned upside down (inverted) so that the displacers are at TDC, and the displacer rods are then marked with a felt pen where they just emerge from the guides.

Always take careful note of where each displacer assembly fits into the cylinder head to avoid mix-ups. The displacers are again removed from the cylinders and each displacer rod is prepared to take its respective con-rod. The rods are first cut ³/₁₆in longer than the felt pen mark, and then slotted ³/₃₂in for ¼in. A ¹/₁₆in hole is drilled laterally in the slotted part of the displacer rod ⅛in from the end (ie, midway in the slotted part). This part will take the con-

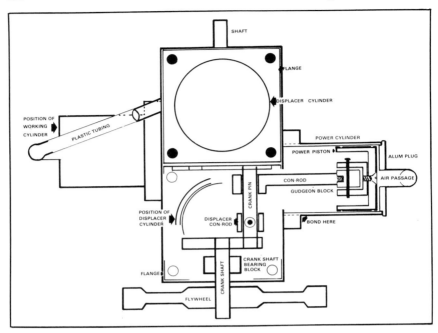

rod and the king-pin or gudgeon pin. The displacer cylinder units are now re-assembled with a thin gasket between the flanges and the cylinder head.

**The power cylinders** are cut from a suitable mild steel tube, in this case from another internal shock absorber tube. These sections have the closed ends and the flanges bonded in place with the 'heat and freeze' method. This operation can be replaced by brazing. The closed ends of the power cylinders are drilled to take ¼ in ID copper bends which will serve as air-vents. The cylinders are then lapped and prepared to take the power pistons. Finally the flanges are drilled for bolting into the engine compartment which is also prepared with four tapped 4 BA holes to take the securing bolts.

**Pistons** may be machined from cast-iron, bms or brass. This engine has two brass pistons machined from solid brass rod with ⅛ in walls and crown. With the pistons in the chuck, a hole is drilled to take the gudgeon block retaining bolt and three oil rings grooved into the piston wall. The pistons are prepared to give a tight sliding fit, the final finish being done by hand after the piston is parted in the lathe.

The gudgeon blocks are machined from ½ in brass or bms, drilled and tapped 4 BA at one end and slotted ⁵/₃₂in to take the con-rods. It is essential that the gudgeon pin is placed as near to the piston crown as possible in order to obtain a long con-rod.

**The multiple crankshaft.** This engine is designed to have the two power pistons firing at the same time. Thus while there is reduced engine speed, there is an increase of power. The crank is fabricated from four webs and three ³/₁₆in silver steel rods.

The webs are cut from ¼ in bright mild steel and are shaped or left square-

**Left** Fig 11.5 *Overall view.*

**Right** Fig 11.6 *Crank webs.*

**Below** Fig 11.7 *Fabricated crankshaft assembly.*

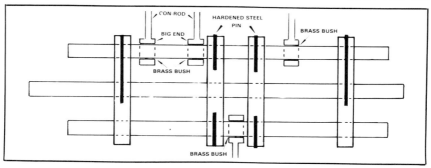

sided as in fig 11.6. The webs are drilled and reamed with three ³/₁₆in holes, ½in between centres. In this operation the webs are secured together to obtain perfect alignment of the bores. One way of doing this is to mark and punch the top web, place and tighten the webs in a drill vice so that there is no movement, and drill and ream the centre hole through the four webs. Secure the webs together with a bolt and then drill and ream the other two holes.

The next stages of the fabricated crank are illustrated in fig 11.7. Three silver steel rods are pushed through the webs, the outer rods being 3½in long while the centre rod is 5in long. Four big-ends, drilled and tapped 4 BA to take the con-rods, are inserted with brass bushes, in the rods before the crankshaft is fully assembled. The position for these big-ends is shown in the illustration (fig 11.7). The rods and webs are lightly bonded in place after aligning the assembly so that the big-ends are directly opposite the displacer rods and the power piston gudgeon blocks.

The next operation is delicate and requires careful handling. The bonded fabricated assembly is placed and tightened in a 3in drill vice and four ¹/₁₆in holes are drilled in the middle and right through the length of the webs. The use of cutting oil is very important during this operation. The assembly is thoroughly cleaned with paraffin, while in the vice, and dried. Later a bonding agent is pushed into the drilled holes, and finally steel pins are hammered through the webs to secure the crankshaft assembly. When the bond has set the surplus pieces of crank are cut off with a fine hacksaw and the sides filed down, leaving the fabricated crank as in fig 11.8. Four 4 BA studs, 1½in long are screwed into the big-ends while small-ends are prepared to fit

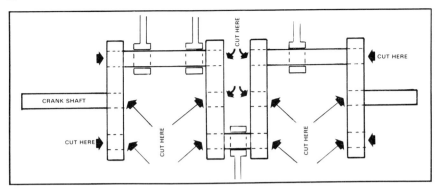

**Above** Fig 11.8 *Completed crankshaft.*

**Below** Fig 11.9 *Burner/cooler.*

into the power gudgeon blocks and the displacer rods. The small-ends are screwed on temporarily to check for con-rod length before the final fitting. The crank assembly is also mounted temporarily in the engine compartment on bearing plates, one at each end, which are bolted from underneath; the assembly is checked for friction-free movement. Brass bushes should be inserted in the con-rod big-ends and in the bearing plates.

Part of the engine, the displacer units, are already assembled. The displacer con-rod small-ends are tried for fitting into the displacer rods with cotter pins and any adjustments to the length of the studs are made with the crank temporarily in place. The power studs are screwed into the big-ends from the lateral holes in the engine compartment, and the small-ends screwed on. The small-ends are then fitted into the gudgeon block and the power cylinders are lightly bolted to the engine compartment. Again any adjustments required to the power con-rod studs are made before the power cylinder bolts are securely tightened.

At this stage all the crankshaft fittings are checked and with the flywheel on, several turns are given to the crank to ensure that there is no gap at the top of the power stroke. When the fitting accuracy has been checked, the crank assembly is tightened and where necessary bonded. A long running-in period is recommended, with the engine coupled to a slow running synchronous electric motor. The air vents are unconnected at this stage. The engine parts have to be well bedded before any attempt is made to start the engine under its own power. Finally the air vents are connected with plastic or rubber tubing.

A 3in × ¾in rim, 12 oz flywheel is recommended for this type of engine. A double ring gas burner, such as described in Chapter 4, will be required. Cooling is by a double water jacket connected to a water tower.

## PERFORMANCE AND MODIFICATIONS

This engine is an easy starter. After an initial period of running-in the engine settles down to provide many hours of trouble-free running. On starting it picks up speed steadily, reaching and maintaining about 600 rpm for long stretches. There is quite a considerable torque for an engine of this size.

The author recommends a modification to the engine compartment, a slightly larger body, ½in higher (ie, 3in internal height) and ⅞in wider (ie, 2½in internal width) with corresponding adjustments in the length of the power piston and displacer con-rods. The increase in length will ease lateral pressure on the bushes and the pistons and reduce friction on the big-end bearings. It is possible to find an extruded aluminium frame section of this size. Alternatively the compartment can be built up from an aluminium sheet formed into a box.

The engine can be made to run with the power pistons firing alternately. This can be done by changing the fabrication of the multiple crank, keeping in mind that the displacer stroke preceeds the power stroke by 90°. This type of crank will amost certainly increase engine speed but there will be a corresponding reduction in engine power.

With minor modifications the engine may be made to work at slightly elevated crankcase pressure. The rear end of the engine compartment may be completely sealed off by fitting a ¼in dural bulkhead, bonded and bolted in place. The crankshaft at this end will need shortening. The front end can be sealed off with a similar bulkhead, using a sealed ball bearing on the crankshaft, backed from inside the engine compartment by a soft rubber washer large enough to cover the ball bearing and held tight in place by a very soft compression spring with an 'O' behind and pressing on the rubber washer to the inner ring of the bearing. This arrangement, or one similar to it, will prevent or lessen air leakage from the crankshaft end.

Pressure into the crankcase is provided, initially, by a handpump attached to a bicycle valve fitted to one side of the engine compartment (seen in fig 11.1). this is an elementary method but, no doubt, the construction of an airtight engine compartment can be vastly improved with experience.

**Engine specifications**

| | |
|---|---|
| Engine compartment | 1¾in external width |
| | 2⅝in external height |
| | 3½in length |
| | 1⅝in internal width |
| | 2½in internal height |
| Dural block | 1¾in width |
| | ⅝in height |
| | 3½in length |
| Displacer cylinders (2) | 1⅜in OD |
| | 1⁵/₁₆in ID |
| | 3¼in internal length |
| Displacer flanges (2) | 1¾in × 1¾in × ³/₁₆in dural |
| Displacers (2) | 1¼in OD |
| | 2¹¹/₁₆in length |
| Displacer rod guides (2) | ½in OD |
| | ³/₁₆in bore |
| | ¹¹/₁₆in length |
| Power cylinder (2) | 1¹/₁₆in OD |
| | 1in ID |
| | 1½in internal length |
| Power flanges (2) | 1½in × 1¼in × ³/₁₆in dural |
| Power pistons (2) | 1in OD |
| | ½in length |
| Displacer stroke | 1in |
| Power stroke | 1in |

CHAPTER 12

# HOW TO CONSTRUCT THE ERICSSON HOT AIR PUMPING ENGINE

This engine (fig 12.1) was designed and built from reproductions found in various publications, mainly from *The Engineer* of that period. This type of engine was originally produced in large quantities and in various sizes, around 1880. It is thought that this is the first type of engine built by John Ericsson which did not incorporate the usual valve system that Ericsson was so fond of. Ironically it is one of the more successful engines he designed. As a pump it became well known for its reliability and performance. The original woodcut reproductions were small in size and could not be relied on for accuracy of measurement and scale. These were enlarged by Xerox to a scale dictated by the availability of materials, and the final designs (fig 12.2), were used in the construction of the model.

The engine is a co-axial type, single cylinder, twin piston (power and displacer), with a series of linkages to the single crankshaft and flywheel. The pump is an external attachment at the other end of the main beam. The first engines built by Ericsson to this design had a single pump but the engine proved to be sufficiently powerful to work two pumps simultaneously and later engines had this modification installed.

The model described here was built on a scale dictated by two main engine components which were readily available at the time — a heavy duty shock absorber outer cover with an internal diameter of 1¾ in, and a flywheel 6¼ in OD, from an old sewing machine. The cylinder was particularly useful because the cylinder was of 'mono' construction (that is solid drawn without any seam welding) and it had an internal mirror-like finish which did not require much lapping. In addition the wall thickness (⅛ in) allowed for brazing without distorting the cylinder and for substantial wall thinning at the hot end to lessen heat conduction.

This model engine can be scaled up or down depending on the availability of materials or on the workshop facilities at the disposal of the modeller. The model can be enlarged sufficiently to perform useful work (pumping hot water), provided that one has access to cheap combustible material or gas. The author has seen a six-inch high model (now in the USA) which works quite efficiently with a fair turn of speed, but without the pumping operation.

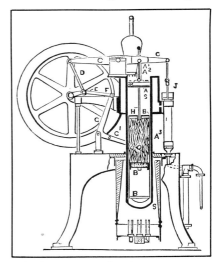

**Left** Fig 12.1 *Completed engine.*

**Above** Fig 12.2 *Drawing of the engine from* The Engineer.

**Left** Fig 12.3 *Engine parts* (The Engineer). *1. Flywheel 2. Crankshaft bracket 3. Crank 4. Connecting rod crank to main beam 5. Main beam 6. Main beam centre bearing 7. Bell-crank link 8. Bell-crank 9. Bell-crank bearing 10. Crosshead assembly 11. Displacer rod 12. Power piston con-rod (2) 13. Pump link 14. Pump 15. Water jacket 16. Main cylinder (above is the water jacket, below the furnace) 17. Main cylinder flange 18. Bed plate 19. Supporting legs 20. Furnace.*

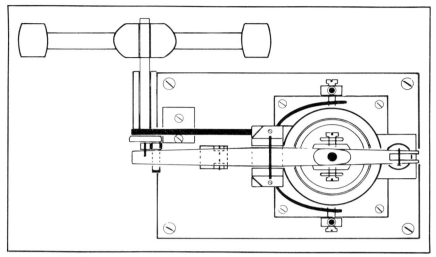

Fig 12.4 *Overall view.*

## CONSTRUCTION

**The platform** is made from ¾ in L-iron bars welded to form a base 6¾ in ×
4⅜ in, and four legs, also from ¾ in L-iron, welded to the base, splayed
slightly outwards for greater stability. The bed-plate, ⅛ in thick, is drilled
2⅛ in to take the cylinder, which is bolted by four 2 BA bolts to the base.

**The linkages** are created to templates cut from the xerox enlargement. The
beam was made from ½ in square bar shaped according to the template. Two
bright mild steel strips, ½ in × ⅛ in and 1in long are brazed sideways to the
beam, midway between the beam centre bearing and the pump link to re-
inforce the beam where an elongated hole is drilled to take the displacer rod.
The connecting rod (beam to crank) can be made adjustable through left and
right threading for fine adjustments to the power piston stroke. Some adjustments may be required during the running-in period.

**The flywheel** The scale requires a 6¼ in flywheel but the use of a slightly
smaller flywheel, up to 5in diameter, will not adversely effect engine
performance.

**The cylinder** Very little work is required on this type of cylinder. A flange,
2¾ in × 2½ in, is brazed (silver brazing is sufficient) to the main cylinder,
about one third of its length from the hot end, to enable the cylinder to be
bolted to the engine base. This method replaces that used by Ericsson who
constructed the cylinder in two parts which were bolted together to the bed-
plate. The water jacket is also (silver) brazed to the main cylinder. This outer
cylinder may be made from thin gauge mild steel or cut from a thin walled
cylinder of suitable size. Two alternative sources for this size of cylinder are
scrap silencer tubes or the outer covers of large shock absorber cylinders.

Work required on the water jacket cover includes the brazing of a ³/₁₆in
mild steel plate in the form of a lip to the top and front of the water jacket.
This takes the main beam bearing and supports the crankshaft bracket. Water

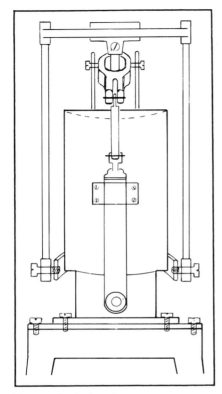

Fig 12.5 *Back elevation.*

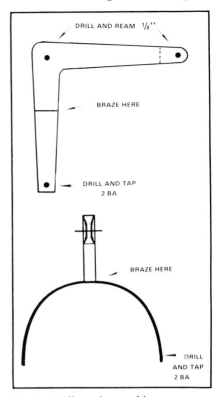

Fig 12.6 *Bell-crank assembly.*

inlet and outlet holes are drilled on opposite sides of the water jacket, the outlet hole being on a slightly higher level than that of the inlet. The inlet hole is elongated or milled sideways to allow for easy flow from the water pump. Four holes are drilled and tapped 8 BA to take the pump housing. The cylinder wall of the hot space is thinned below the flange or bed plate to ensure heating efficiency and to reduce heat conduction through the length of the cylinder. This operation requires the use of a heavy chuck capable of taking a 2⅛ in cylinder.

**The crankshaft bracket** which takes the crankshaft bush (brazed or threaded), is bolted to the bed plate and also, in this model, to the lip of the main cylinder water jacket. In Ericcson engines, the bracket was bolted direct to the water jacket casing; apart from supporting the crankshaft in earlier engines it also supported the bell-crank shaft. In a model engine of modest size, since the bracket is made out of ⅛ in bright mild steel or ¼ in dural plate, it is good policy to separate the crankshaft support from the bell-crank support, and to provide for an alternative bracket for the bell-crank. In practice this crank and its linkage absorb a substantial amount of impact with each movement of the displacer, particularly if this is heavy (see modifications discussed at the end of the chapter). The construction of a good bracket for the bell-crank can be seen in figs. 12.3 and 12.11. The bracket is bolted to the bedplate, the bolt holes drilled and tapped 2 BA, in the direction of the movement of the crank.

**The bell-crank** unit (fig 12.6) is made up of two separate parts, brazed (or bolted) together: an L-shaped arm which pivots at its centre, and a horseshoe bracket. The vertical part of the L-arm is linked to the crank (and the flywheel), while the horizontal part, brazed to the horseshoe bracket, is part of the leverage which lifts and lowers the displacer via the crosshead assembly. The L-arm is cut from ¼ in dural or from ³/₁₆ in mild steel. An alternative method, used in this model, is to form the arm from two pieces of mild steel flat bar, ⅝ in × ¼ in, brazed in an L-shape and filed down to give the right outline. The horseshoe bracket is shaped from mild steel strip, ½ in × ⅛ in. Since this bracket carries the weight of the crosshead and the displacer cylinder, the bell-crank assembly should be quite rigid.

**The crosshead assembly** consists of the following parts: four BMS blocks ⅝ in × ½ in × ¼ in thick; two silver steel rods, 6¼ in (5¾ in between threads), ³/₁₆ in OD; T-piece cut from brass or mild steel bar, 1¼ in × 1 in × ½ in; one silver steel cross shaft 3¾ in between threads, ³/₁₆ in OD. The construction of this assembly is fairly straightforward and can be followed easily from figs 12.1 and 12.5.

The four mild steel blocks are cut from a flat bar ½ in wide and ¼ in thick. Two are drilled and tapped 2 BA in the centre and again midway along the edge of the ½ in wide sides. Two other blocks are drilled ³/₁₆ in in the centre, and drilled and tapped 2 BA midway along the edge of one of the ½ in wide sides (fig 12.5). The two silver steel rods are threaded 2 BA at both ends for ¼ in, leaving a length of 5¾ in between threads. The silver steel shaft, cut from a ³/₁₆ in silver steel rod, is similarly threaded at both ends. The T-piece is drawn and then cut from a brass or mild steel bar (as in fig 12.7). a ³/₁₆ in hole is drilled and reamed through the top of the T-piece and then a hole is drilled and tapped 2 BA to take the displacer rod when the crosshead is assembled on the engine.

The assembly of the crosshead takes the following order. The T-piece is mounted on the cross shaft, and the two mild steel blocks are screwed into the shaft, ensuring that the 2 BA tapped holes are pointing in the same direction. This part of the assembly should be stiff, with only the T-piece having free movements. The silver steel rods are then screwed into the blocks and

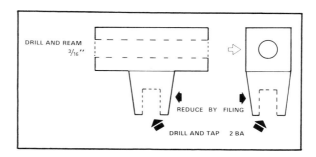

Fig 12.7 *T-piece.*

tightened. Finally the remaining two blocks are screwed in tight, parallel to the top two blocks. The completed assembly should be quite rigid when mounted by means of two 2 BA bolts to the horseshoe bracket on the bell crank.

**The power piston** is machined from mild steel bar, or fabricated from mild steel cylinder brazed at the closed end with a $^3/_{16}$in disc. Machining the piston is a multi-stage operation, first reducing the diameter to obtain a tight fit inside the cylinder, then cutting two grooves for 'O' rings or four shallow grooves for oil retention. The crown is faced, centre-drilled, drilled and tapped $^3/_8$in BSF. Finally the piston is parted off while in the chuck. The piston is then reversed in the chuck and faced internally with an end-mill to give a good seating to the displacer guide bush. The last part of the operation is lapping and fitting the piston, ensuring a good sliding fit in the cylinder.

**Displacer assembly** The first displacer was made from the outer cover of a shock absorber. However, the engine jumped all over the work bench and the thumps could be heard from a distance. An immediate modification was made and is explained here.

The displacer can be made from an aluminium lightweight deodorant container. This size of container usually has a concave bottom and this construction can be utilised if the curvature is reversed by lightly hammering the bottom internally on a soft piece of lead. An aluminium plug is machined, centre-drilled, drilled and tapped 4 BA to take a $^3/_{16}$in silver steel rod similarly threaded. The plug is bonded in the displacer ensuring that the rod is perfectly aligned to prevent friction against the cylinder wall.

**The displacer rod guide bush** which takes the displacer rod, is another multi-stage operation at one go. A 1in mild steel bar is faced for 1¾in. The first ¼in is reduced and threaded $^3/_8$in BSF, with a slight undercut at the end of the thread. A collar, ½in long, with an OD of ¾in is then machined, while the remaining length is machined to ½in OD. The guide is centre-drilled, drilled and reamed $^3/_{16}$in and parted off. The bush is then bench-fitted and prepared to take the twin con-rods which link the power piston to the main beam. The ¾in OD collar is filed down on opposite sides of the circumference by $^1/_{16}$in. The centre of each filed area is drilled and tapped 2BA to take king-pin bolts. The con-rods are screwed into the collar with a light application of Loctite Screwlock 222 to the bolts and the bush is screwed into the power piston, with a fibre washer, and well tightened.

**The furnace (burner or heater)** (fig 12.8). The model Ericsson engine looks like the real engine if a furnace cover is added. The furnace cover and the furnace shield help also in keeping the flames concentrated on the hot space of the cylinder, whatever type of burner is used. The shield is made from thin sheet metal in the shape of a funnel with the top narrow end cut off. A number of $^3/_{16}$in air vents are drilled along the top edge as an aid to convection.

The furnace cover is formed from sheet metal or cut from a large diameter cylinder. In this model, a 3in OD silencer tube 3¾in long is used. Four lugs are brazed to the cylinder and drilled to take 2 BA bolts. The cover is bolted to the engine frame and bed plate from underneath. The furnace bottom is open but a wide wire mesh may be used as a grate capable of taking solid fuel

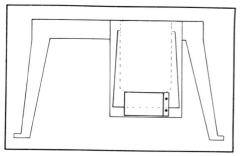

**Above** Fig 12.8 *Furnace.*

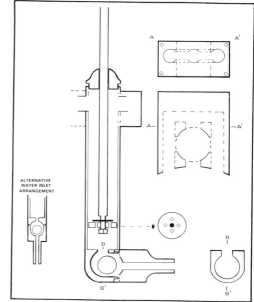

**Right** Fig 12.9 *Pump assembly.*

**Right** Fig 12.10 *Plunger assembly.*

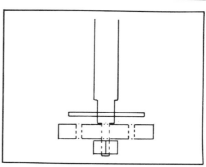

cubes or a gas burner. A furnace door is constructed by cutting an opening in the side of the (furnace) cover, 2in long and 1in high. A hasp, suitable cut and rounded, makes a fine door with the narrow hinged arm bolted to the furnace cover.

**The pump** (fig 12.9) consists of a brass barrel through which a plunger is pushed, allowing water to flow through. When the plunger is pulled up, water above the plunger is emptied through bores in the pump box into the aperture or water inlet in the water jacket cover. The plunger consists of a brass washer into which are bored four $^3/_{64}$in holes. On top of the bored brass washer rests a thin brass washer which completely covers the four holes and which is allowed a movement of $^1/_{16}$in along a machined $^1/_{8}$in shaft. The pump inlet is a separate brass piece which is fine threaded and screwed into the main barrel. A plastic ball serves as a valve, alternately closing and opening one or other of the water passages (fig 12.9).

## PERFORMANCE

The author's model Ericsson engine worked at the first attempt. Initially the engine required a substantial amount of heating, the thickness of the cylinder wall being a contributing factor, until the wall was thinned down by $1/16$in along the bottom third of its length. The engine still requires a fair amount of initial heating until it starts to pick up speed, but once an even and rhythmic cycle is achieved, only minimum heat is required. A no-load speed of 450 rpm can be achieved with the pump disconnected and with a lightweight displacer. Under pump load the speed drops to 200 rpm provided that the water container is at about table top level.

Cooling is absolutely essential and the first Ericsson engines provided for this by using the pumped water to cool the jacket before being utilised for irrigation etc. Later models had two pumps working at the same time, a double-acting pump for deep wells and a single-acting pump for shallow wells or low-level tanks. It appears that the single-acting pump discharged water through the cooling jacket of the engine for domestic use, while the double-acting pump discharged water direct for irrigation and other uses.

Two modifications to the model are suggested by the author. If the displacer is slightly weighty, a counterweight on the bellcrank will largely equalise the weight of the displacer and will produce a less powerful thumping action and smoother movement of the drive mechanism. The counterweight may be increased or decreased until the best result is achieved. Pump efficiency may suffer as a result of the use of the counterweight, but only slightly.

The other modification, mentioned before in passing, is for an adjustable connecting rod (beam to crank). After a number of attempts to get the right stroke, as a result of working to rough measurements, and each attempt involving the cutting and drilling of a new con-rod, it was decided to machine a con-rod with left and right threads, similar to the bottle-screws used on yachts. A left-hand tap and die set is necessary; this can be purchased from engineering supply firms (some advertise in *Model Engineer*). This method of adjusting the stroke is very effective and will reduce dead space between the power piston and the displacer. (This adjustable con-rod can be seen in fig 12.1.)

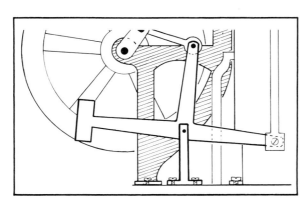

Fig 12.11 *Counterweight to bell crank.*

CHAPTER 13

# HOW TO CONSTRUCT 'PROVA II', A COMPETITION TYPE CO-AXIAL STIRLING ENGINE

'Prova II' (winner of the Brian Thomas Memorial Trophy, 1984 Model Engineer Exhibition, Wembley) is a test bed engine designed for experiments on external annular regenerators (see Chapter 3). The concept is for an engine which allows the regenerative matrix to be changed fairly easily without disturbing the engine mechanism. The design therefore incorporates a number of important features; A main or working cylinder that can be taken out and modified without removing the power piston/displacer assembly; a regenerative matrix that can be enclosed between the working cylinder and the regenerator cover, the whole assembly being gas tight; and thirdly a gap

Fig 13.1 *Completed engine.*

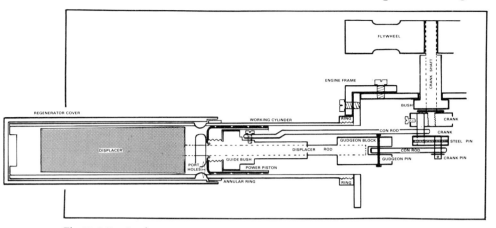

Fig 13.2 *Engine layout.*

between the working cylinder and the regenerator cover which can be enlarged as required to accommodate larger matrices. The engine design also incorporates a construction feature which consists of an easily dismantled engine frame so that any part of the engine can be removed without disturbing the rest.

The materials for the engine can be obtained from scrap or surplus supplies with the possible exception of the stainless steel shim (regenerative matrix), the silver steel rods and the flywheel. The outer regenerator cover, the retaining ring and collar, the working cylinder and the displacer cylinder of the winning engine were all obtained from the internal cylinders of various different sizes of shock absorbers.

## GENERAL ENGINE LAYOUT

The working cylinder is open-ended, the front part threaded to fit into a ring brazed on a flange. The rear end, apart from being open is partly filed down to allow the uninterrupted flow of gas during the expansion stage. Four ports are milled (or drilled) about midway along the length of the cylinder. A regenerator cover is machined to fit on to the working cylinder, long enough to cover the port holes, while leaving a narrow gap. The working gas moves through the rear opening, through the regenerator gap, through the port holes back into the main cylinder. The assembled cylinder is mounted onto an engine frame which allows for stroke and timing adjustments with comparative ease.

## METHOD OF CONSTRUCTION

**The main or working cylinder.** The front end of the main cylinder is threaded 32 tpi for ¼ in but other fine threads would also be suitable. Alternatively the cylinder may be brazed to a flange. The back end is segmented by fine saw cuts into eight equal sections, alternate sections filed down by ⅛ in. Four portholes are machined in the cylinder, their position being at the place

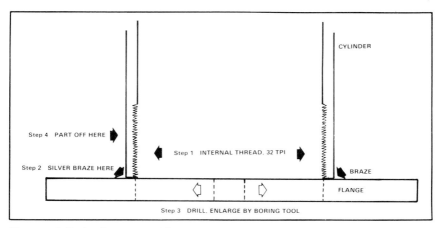

Fig 13.3 *Cylinder flange assembly.*

where the power piston reaches TDC. The portholes are milled, with the use of a dividing head and chuck to give four points where the end mill, $^3/_{16}$in, makes its first cut. Each cut is slowly lengthened sideways to give four portholes of $^3/_{16}$in × $^5/_8$in. Alternatively, 12 holes, $^3/_{16}$in may be drilled, equally spaced around the circumference. Care is taken not to damage or to twist the cylinder. Finally a boring tool is used to remove any sharp metal edges left by the milling/drilling operation.

The final stage of construction on the cylinder involves the mounting of a ring just forward ($^1/_{32}$in) of the portholes. This ring, which is tight fitting, is lightly bonded in place. It is important to ensure that the OD of the ring is slightly over the unmachined ID of the regenerator cover. In this engine the ring OD is 1$^1/_8$in while the regenerator cover ID is 1$^1/_{32}$in.

The next operation is to cut a female thread inside a cylinder of relative size, (OD 1$^1/_{16}$in) to fit the external thread of the main cylinder. The tube is then silver brazed, thread down, on to the brass plate. The threaded cylinder is placed in a chuck, with the brazed-on plate (or flange) facing out. By drilling a series of extending holes and finally using a boring tool, the bore is sufficiently enlarged to enable the insertion of the displacer and power piston without difficulty. While the work is in the chuck, the brass flange is checked for face accuracy. After this check ¼in of the threaded cylinder is parted off, and with a light stroke from a fine half-round file, the working cylinder is screwed in (see fig 13.3).

**The regenerator cover** (fig 13.4) Two stages of operation are required here — welding with oxyacetylene and iron filling rod, and the use of a boring tool to widen the first ¼in length of this cylinder to 1$^1/_8$in to give a nice sliding fit on the ring which has been bonded to the main cylinder. The regenerator cover may be externally thinned down for two-thirds of its length to lessen heat conduction.

**The displacer piston** construction requires more attention than any other part of the engine, probably even more than the power piston itself. A thin-

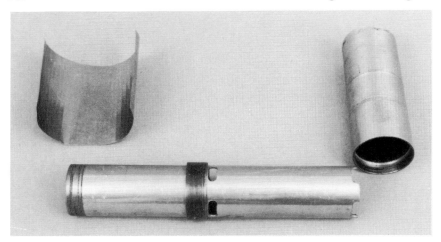

Fig 13.4 *Regenerator cover and shim.*

walled cylinder is found or machined to obtain a near sliding fit in the
working cylinder. This is brazed at one end using thin gauge sheet metal and
carefully avoiding distortion. In the second stage a mandrel is used — a tight
fitting hard-wood dowel is sufficient for this purpose. The cylinder is held in
place at the outer end by a revolving centre after the closed end is centre-
drilled. This hole can be brazed later but before an aluminium plug is used to
seal the other end.

The third stage is to machine the displacer cylinder into a slight tapering
shape towards the brazed end, starting the taper just ¼ in from the end
plugged by the mandrel. The front end of the displacer (the widest part)
serves as a piston to 'encourage' the working gas to move into the outer
chamber enclosed by the regenerator cover. The cylinder can be left
untouched in external diameter, but if the fit is too close, no matter how well
centred a displacer is, it is sure to scrape along the cylinder wall after a few
turns of the engine.

A longish displacer is necessary to provide for a long passage of the working
gas through the external regenerative matrix. Therefore in order to strike a
balance between a long displacer and the avoidance of friction between the
displacer and the cylinder wall, the displacer should be tapered. The trick of
course is to achieve this with an already thin-walled cylinder without
removing the brazing or ruining it completely. The displacer rod is threaded 4
BA and inserted into the aluminium plug. Care is taken to align the displacer
and the rod correctly in order to avoid any scratching while it is moving inside
the working cylinder.

The last job on the displacer assembly is the gudgeon block/con-rod fitting
(fig 13.5). There are several ways of making a gudgeon block — the simplest
type is made from a ⅜ in × 1 in brass bar, drilled ³/₁₆in to a depth of about
½ in. The other end is slotted ⅛ in × ½ in deep, and prepared to take the
con-rod. In line with the slot and about ³/₁₆in from the bored end, a hole is

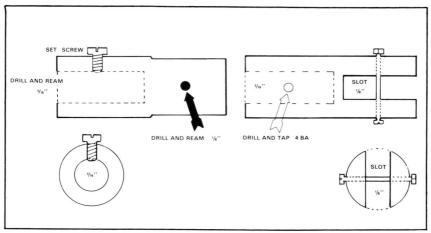

**Above** Fig 13.5 *Gudgeon block.*

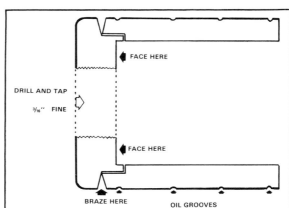

**Right** Fig 13.6 *Power piston.*

drilled and tapped 4 BA to take a set screw. The gudgeon block is then drilled ¼ in to take the gudgeon pin across the middle of the slot. The con-rod should fit nicely without friction or binding. The gudgeon block gives an adjustment of about ¼ in, that is ⅛ in each way.

The power piston is constructed from cast iron or machined from suitable bright mild steel bar (fig 13.6). Work on the piston is a multi-stage operation, first reducing the diameter to obtain an almost perfect fit, cutting four oil grooves and finishing off the top (crown) in a fairly rounded end — required in view of the portholes arrangement. While still in the chuck, the closed end is drilled and tapped ⅜ in BSF and then parted off with about ¾ in of extra length. In the final stage the piston is reversed in the chuck and faced internally with an end mill to ensure a tight fit when the displacer rod guide bush is screwed in. The piston is then parted off, ready for the lapping operation. The power piston used in Prova II was fabricated from a cast iron pipe brazed to a $^3/_{16}$in mild steel disc, as shown in fig 13.6.

The displacer rod guide bush (fig 13.7) The machining of this guide is a

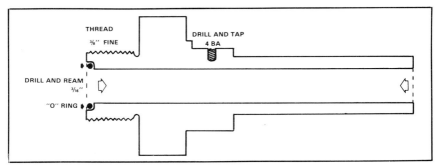

Fig 13.7 *Displacer rod guide bush.*

four stage operation at one go. A ¾ in mild steel bar is faced along its length. The first ⅜ in is reduced and threaded ⅜ in BSF with a slight undercut at the end of the thread. A collar of about ¼ in is left untouched (¾ in OD), a second collar is machined to ½ in OD while the remaining length is reduced to ⁵/₁₆in OD. The guide is first centre-drilled, then drilled and reamed ³/₁₆in and parted off.

The next operation is a bench one. The second collar (½ in OD) is filed on one side to remove about ¹/₁₆ of its diameter; a hole is then drilled in the centre of this part and tapped 4 BA to take a bolt which will be the kingpin of the lower piston con-rod. Once the power piston con-rod has been prepared, it is bolted to the guide bush, allowing it free movement. The guide bush is then assembled into the power piston with a fibre washer and a smear of Loctite Screwlock 222, and the whole assembly tightened.

**The engine frame** consists of two plates, a cylinder plate and a crankshaft bearing plate which are bolted to a bracket in turn screwed to a base. This type of construction allows for either plate to be removed without the necessity of unscrewing the other. The plates are cut and shaped from ⅛ in to ³/₁₆in brass or bright mild steel, 2½ in long and 2in wide. The centre point of the cylinder ring is 1½ in from the L-corner, while the centre point of the crankshaft bush is 1¾ in from the L-corner. Two important requirements should be kept in mind. The first that the plates must be at 90° to each other and the other that the centre points of the cylinder ring and the crankshaft bush are perfectly aligned.

**The crankshaft bush**, 1½ in long, is machined from mild steel bar ¾ in diameter. Machining, on the lathe, is a multi-stage operation at one-go. The bush is faced, then 1¼ in length is reduced to ½ in. The bush is centre drilled, and then drilled and reamed ³/₁₆in. One or two oil holes may be drilled in the bush for lubrication. The bush is fitted to the side plate and sweat soldered or silver brazed. It is advisable to re-finish the bore with a reamer if any heat is applied to the bush.

**The lapping and fitting of the power piston** (see Chapter 8) is completed before any further assembly work is undertaken. At this stage it is required to provide the power piston with a stroke of ⅜ in; to provide the displacer with a stroke of ⅞ in; and to provide a variable phase angle of between 90° and 110°.

A ³/₁₆in shaft is machined and threaded 4 BA to take the power crank, made from a bright mild steel square section ¼ in × ¼ in × ¾ in, marked at the centre and ³/₁₆in from the centre. The centre hole is drilled and tapped 4 BA while the other hole is drilled ⅛ in clear. The crank is then drilled laterally and tapped 4 BA on the side of the ⅛ in hole.

Another piece of mild steel section, ³/₁₆in × ³/₁₆in × ¾ in is then prepared for the second crank (for the displacer stroke). Two holes, ½ in apart, are marked and drilled ⅛ in clear, into which are inserted two pins to a tight fit and preferably bonded as well — on opposite sides to each other. Once in place, a ¹/₁₆in hole is drilled laterally through the length of the crank to secure the pins, preventing movement by means of a steel pin. Crank assembly is shown in fig 13.8. It is advisable to thread the pin designed to take the displacer con-rod in order that a lock nut can be used should there be a need to prevent the con-rod from slipping out. Alternatively a split pin/washer arrangement can be devised. With the two cranks completed, it remains only for the stroke and timing mechanism to be fitted and set.

Once the power piston has been lapped, the first part of the stroke and timing operation is undertaken, fitting the power piston con-rod to the power crank. This is done by first pushing in the power piston as far as it should go, that is to the edge of the port holes. The crank is then turned so that the drilled ⅛ in hole is nearest to the power piston; in this position, the power piston is at TDC. A mark is made on the con-rod which is then drilled ⅛ in to take the crankpin. Once fitted the crank should be turned several times to ensure that there is no binding.

**Displacer fitting** In this part of the operation the displacer is first inserted in the working cylinder and the gudgeon block fitted on the displacer rod. As explained in Chapter 6 the closest the displacer is to the power piston during

Fig 13.8 *Crank assembly.*

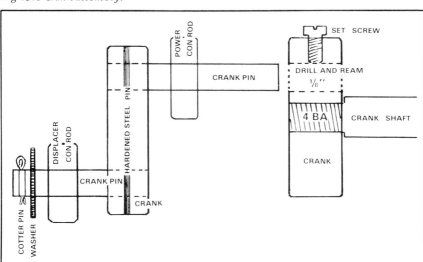

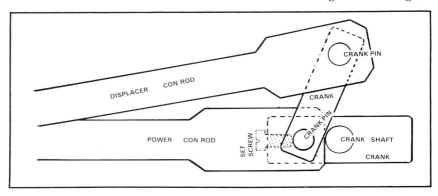

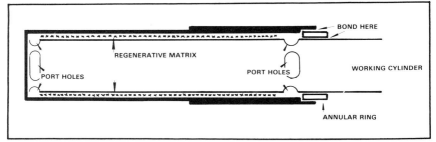

**Top** Fig 13.9 *Drive mechanism.*

**Above** Fig 13.10 *Regenerator cover.*

its movements in the cycle is when the cranks are at approximately 10 past 11, that is the piston con-rod is at 5 mins to the hour, while the displacer con-rod/crank-pin is at 10 mins past the hour. Figure 6.20 explains this position clearly.

The space between the power piston and the displacer should be at its minimum in this position; this is the starting point of the engine setting. Any small adjustment can be made later from the gudgeon block. Major adjustments on the other hand involve either the length of the displacer rod or the length of the con-rod. In this position (10 past 11), the cranks are approximately at 90° to each other, and therefore marking and drilling for the hole where the crank pin should fit into the·displacer con-rod should be made here. Throughout the fitting of the power piston and the displacer the working cylinder is uncovered so that all the operation is subject to visual correction — there is therefore no room for miscalculations or incorrect spacing.

**The regenerator matrix** The first experimental matrix is made from 0.002in stainless steel shim, similar to that used by Robin Robbins and Andy Ross in their engines. The steel shim is ruled by a scriber every ⅛in along the path of the working gas — these lines cause the shim to corrugate and to bend naturally along the circumference of the cylinder. This construction allows the working gas to pass on both sides of the shim, thus doubling the surface of the matrix exposed to the working gas. The gap between the regenerator (outer)

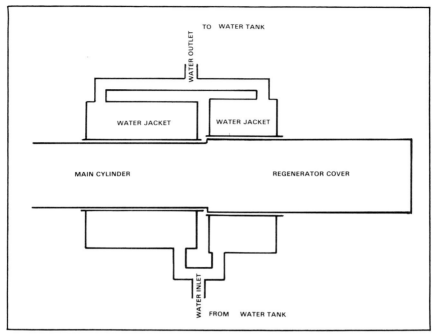

Fig 13.11 *Cooler assembly.*

cover and the working cylinder is of $1/64$in just barely enough to accommodate the corrugated shim while at the same time limiting dead space.

**The final stages of assembly** with the power piston and the displacer in position and the mechanism fully adjusted, involve the positioning of the steel shim between the outer cover and the working cylinder. Care is taken that the matrix does not slip forward to cover the portholes and some juggling with the shim may be necessary to ensure correct positioning. A smear of bonding agent around the annular ring should hold the regenerator cover quite fast without allowing any gas leakage. Contrary to what one expects, the bond is not affected by the heat from the burner once the cooler is slipped over that area.

The best type of cooling is obtained by an assembly of two water jackets, one for the regenerator cover and one for the power piston area. Since there is a difference in the diameters of the two areas to be cooled, it is difficult to construct one cooling jacket for both. The alternative is a combination of a water jacket for the regenerator and a fin cooler for the power area. This second arrangement is not as efficient as the first. One should not rely on cooling the regenerator cover only, since the working cylinder becomes quite hot and increases the possibility of the engine seizing.

The water jackets are connected by plastic or rubber tubing to a cooling tower to provide for thermo syphon action (see Chapter 4). If fin coolers are used, they should have sufficient heat transfer area to eliminate as much

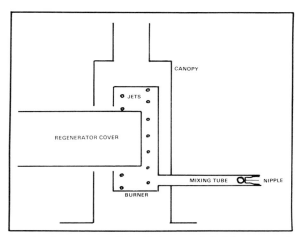

Fig 13.12 *Burner assembly.*

surplus heat as possible. It is not advisable to run the engine for long periods with such an arrangement. With small well constructed water jackets this engine can be quite attractive in appearance.

The burner consists of an enclosed ring with 24 jets of $1/16$in diameter; the mixing tube holds a Camping Gaz burner jet while a copper pipe connects the burner to a Camping Gaz blowlamp through a special adaptor fitted for the purpose. The engine responds quickest to heat if the burner is positioned at the extreme end of the cylinder such that the last row of jets throw a curtain of flames to the rear of the cylinder, while the first row of jets heat the rear $1/4$ in of the regenerator cover. In this way the working gas is forced to make constant contact with the closed end of the regenerator cover, picking up more heat and more quickly.

**The use of 'O' rings** should not be necessary as an aid to compression in a good gas-tight power piston but if for some reason the fitting is suspect, the use of such a ring or one of a PTFE twisted strand in the bottom oil groove may give the engine better performance. On the other hand an 'O' ring at the top of the displacer rod guide can help in two respects — by preventing gas escape via the displacer rod and by preventing any lubrication oil applied to the displacer rod from contaminating the regenerative matrix.

**Running-in the engine** is advised before the regenerator cover is bonded to the working cylinder. It is advisable to give the engine a long running-in by coupling a pulley on the crankshaft (in place of the flywheel) to a slow synchronous motor, leaving this on for some hours with an occasional drop of very fine oil (see Chapter 8). When the regenerator cover has been fitted and the bond set, another but shorter running-in is given to the engine. The final stage of running-in comes when the engine has consented to run. In the initial stages care is taken to run the engine for short periods at slow speed, increasing both the speed and duration over a number of runs.

## PERFORMANCE

The first run of 'Prova II' under power clocked a no-load speed of 1,200 rpm. Subsequent runs, after a slight adjustment to the phase mechanism by increasing the angle from 100 to 110°, gave an increase of speed to a steady 1,830 rpm. Two other results are evident. This engine takes longer to start than other co-axial engines due to the particular regenerator cover construction. The other point is that on turning off the heat the engine takes a longer time to stop, this is due to the type of regeneration. Obviously other types of regenerative matrices may be experimented with. Stainless steel mesh gauze, stainless steel wire, fine glass rods or ceramic strips are all possibilities.

## Engine specifications

| | |
|---|---|
| Working cylinder | 1in OD |
| | $^{29}/_{32}$in ID |
| | 6in length |
| Outer cylinder | 1$^3/_{16}$in OD |
| (regenerator cover) | 1$^1/_{32}$in ID |
| | 3½in length |
| Power piston | $^{29}/_{32}$in OD |
| | 1$^1/_{16}$in length |
| | ⅜in stroke |
| Displacer | $^{57}/_{64}$in (approx) OD |
| | 2$^{17}/_{32}$in length |
| | $^{15}/_{16}$in stroke |
| Portholes | 4 of $^3/_{16}$in × ⅝in |
| Crankshaft & displacer rod | $^3/_{16}$in silver steel |
| Annular ring | 1⅛in OD |
| | 1in ID |
| Displacer rod guide bush | ¾in OD |
| | 1¾in length |
| | $^3/_{16}$in bore ID |
| Cranks: Power | ¼in × ¼in × ¾in |
| | $^3/_{16}$in between centres |
| Displacer | $^3/_{16}$in × $^3/_{16}$in × ¾in |
| | ½in between centres |
| Gudgeon block | ⅜in OD |
| | 1in length |

CHAPTER 14

# HOW TO CONSTRUCT 'SUNSPOT', A SOLAR-POWERED STIRLING ENGINE

'Sunspot' is a small co-axial solar-powered Stirling engine (fig 14.1). Heating the engine is by means of the concentration of the sun's rays on the hot end of the cylinder with the aid of a parabolic reflector aimed at the sun. There are some disadvantages to this type of heating, mainly because of the total dependence on bright sunlight and the need to aim the reflector directly into the sun but these disadvantages are outweighed by the satisfaction of using free energy. In fact this method of obtaining mechanical energy is one of the most direct ways of tapping the power of the sun.

The use of a parabolic mirror or reflector is not the only method of using the sun's rays to power a Stirling engine; an alternative method is the use of the Fresnel lens which also concentrates the rays on to a focal point (fig 14.2). The main difference between the parabolic reflector and the Fresnel lens is

Fig 14.1 *Completed engine.*

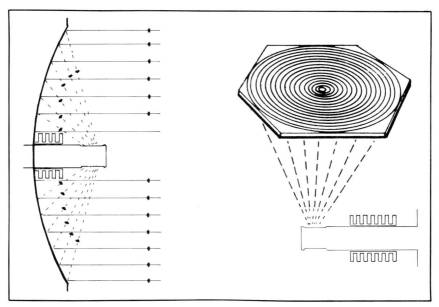

Fig 14.2 *How the parabolic reflector and Fresnel lens work.*

that the parabolic reflector heats the circumference of the hot end of a cylinder while the Fresnel lens heats one particular area or point, although the temperature at this spot can be higher. Another difference is that the cylinder with its hot end heated by the reflector is almost always perpendicular or nearly so, whereas an engine heated by the Fresnel lens is almost always horizontal. The engine described in this chapter can be made to work with other common fuels and is not entirely dependant on the sun's energy. It works just as easily with gas, or methylated spirit. The engine can be made to stand in any position to work with the fuel or energy chosen.

The construction of this engine is only slightly different fromthat of 'Prova II' described in Chapter 13 and of 'Dyna' in Chapter 15, the difference being in the crank assembly. The materials used in the building of the original 'Sunspot' were mainly from scrap or surplus supplies involving minimum expenditure. No doubt luck played a great part with the finding in a scrap heap of a parabolic reflector, part of some hospital or clinic equipment, which was in reasonably good shape and condition (fig 14.3). The reflector was chromed for a small sum by a private firm specialising in this work. The most important part of this particular project is the reflector. It should be perfectly parabolic in shape so that maximum concentration of the sun's rays can be focused on the hot end of the working cylinder. A parabolic mirror or reflector has the property of bringing the rays parallel to the axis to a point of focus.

The greater the circumference of the reflector, the greater the concentration of the sun's rays. A minimum circumference of 12 in diameter is required although a diameter of between 15 in to 18 in is more desirable. In order to

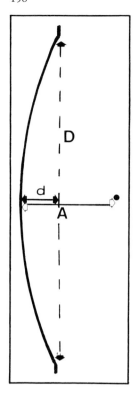

**Above** Fig 14.3 *Scrap reflector—original state.*

**Left** Fig 14.4 *Focal point of parabolic reflector.*

find the focal point of a parabolic reflector the following formula is used:

$$A = \frac{D^2}{8d}$$

Where A is the focal length, D is the diameter of the reflector, and d is the depth at the centre of the reflector when a straight line is drawn across the diameter.

### ENGINE CONSTRUCTION

The reflector used in 'Sunspot' has a focal point of 3in, the cylinder length is 5½in, of which 3¼in is within the reflector sphere. A fin cooler is machined and mounted directly in front of the reflector base, while a similar but smaller fin cooler is mounted behind the reflector. The two fin coolers, apart from their use in cooling the cylinder, are also used to retain the reflector rigidly in place. The cylinder is ring-mounted on an engine frame which holds a multiple crank and supports a flywheel. A light alloy displacer and a cast iron piston are used for the working mechanism.

**The cylinder** is machined from bright mild steel, (ex-shock absorber internal cylinder), brazed with iron filing rod at the closed end. Careful machining on the lathe is essential to the success of the project since the collection of sufficient heat at the hot end is as much dependent on the reflector as it is on the cylinder wall thickness. The most important part of machining is the reduction of the wall thickness of the hot end of the cylinder for a length of ¾ in from the brazed end, ensuring that the brazing is not touched or removed, that the cylinder is not distorted and that the wall is almost paper thin.

A further 2¼ in length of the cylinder is slightly reduced to prevent heat conduction along the length of the cylinder covered by the displacer and the displacer stroke. At the other end the cylinder is threaded 32 tpi for ¼ in. Alternatively a flange may be brazed on the cylinder, care being taken to avoid distortion. Any internal boring and honing should be completed at this stage, after extensive external lathe work has been undertaken.

**The engine frame** There are different ways of constructing the engine frame. The method described in Chapter 13 is an excellent alternative to the one described here. So, for that matter, is the drive mechanism which can be adapted to this engine.

**Right** Fig 14.5 *Engine in construction.*

**Below** Fig 14.6 *Overall view.*

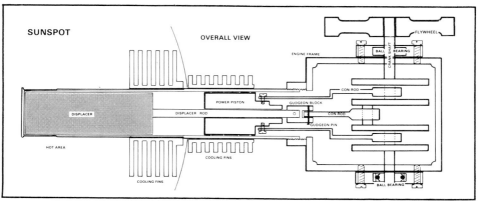

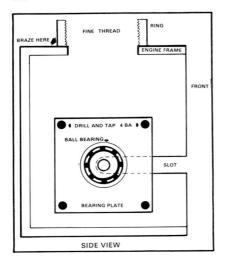

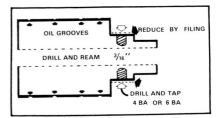

**Left** Fig 14.7 *Engine frame and bearing plate (side view).*

**Above** Fig 14.8 *Power piston.*

**Right** Fig 14.9 *Fabricated crankshaft assembly.*

The engine frame illustrated here (fig 14.5) is taken from surplus electrical equipment which consisted of a number of brass rectangular boxes as part of its construction. Each box, 2½ in × 2¼ in × 2¼ in deep, has its narrow sides bored ¾ in. A ring, bored and threaded internally 32 tpi to take the main cylinder, is silver brazed around a bore, this arrangement allowing for con-rods to be fitted on to the cranks housed within the frame. One minor modification to this frame is necessary in view of the need for the crank assembly to slide into its place within the engine frame. Two holes, ⁷/₃₂in, are drilled in the sides of the frame and slots cut with a fine hacksaw blade (fig 14.7).

Two Dural plates, 1¼ in × 1¼ in × ¼ in are drilled ⅝ in to take ball bearings for the crankshaft, while four holes, ³/₁₆in, are drilled at the four corners to secure the plates to the sides of the frame when the crankshaft is mounted. This assembly allows for the crankshaft, bearings and con-rods to slide out of the frame without difficulty when any adjustment is required. Alternative engine frames may be constructed to allow for adjustments as necessary. A typical engine can have one or both sides bolted to a cylinder flange with supporting brackets.

**The displacer/regenerator** is best constructed from thin-walled bright mild steel tubing brazed at one end and sealed at the other end with an aluminium plug, drilled and tapped 4 BA to take ⁵/₃₂in silver steel rod. Sunspot has an alloy displacer (ex-perfume container), whose regenerative effect is barely adequate. To complement the displacer regeneration a .002in stainless steel shim, raised on both sides by dimples, is inserted to press against the cylinder wall and reduce the annular gap.

**The power piston** (fig 14.8) is machined from cast iron and is a lathe operation of several stages at one-go. A typical construction method is as follows. A 2in length of 1in OD cast iron is faced and reduced in stages; a length of 1¼ in is reduced to ⅞ in OD to obtain a good piston fit, a collar of ¼ in length reduced to ⁹/₁₆in OD, and a length of ½ in reduced to ⅜ in OD.

## FABRICATED CRANKSHAFT ASSEMBLY

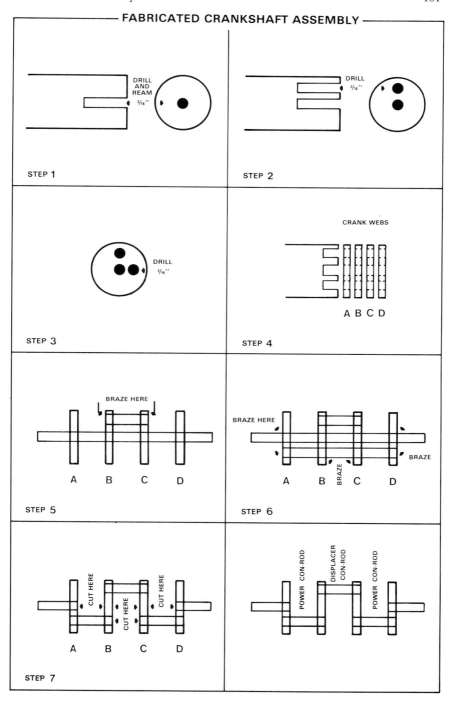

While in the lathe the power piston assembly is centre-drilled, drilled and reamed $^5$/$_{32}$in ($^3$/$_{16}$in bore is a suitable alternative), and finally parted off. The piston is lapped at this stage. The $^9$/$_{16}$in OD collar is filed $^1$/$_{32}$in on opposite sides and prepared to take two con-rods. Two holes, drilled and tapped 4 BA will take the wrist pins for the con-rods.

**The crank** is constructed from four discs or webs, with crank pins fitted in the discs and silver brazed in place. The construction (fig 14.9), although delicate in execution, is quite straight forward provided that the steps explained here are followed in the right order. **1** A bright mild steel bar, 1½ in OD, is faced and prepared in the lathe. It is centre-drilled, then drilled $^3$/$_{16}$in to a depth of 1in. **2** The bar is secured exactly vertical in a drill vice and marked $^5$/$_{16}$in from the centre. The bar is then drilled $^3$/$_{16}$in at this mark, to a depth of 1in. **3** The bar is then marked $^{25}$/$_{32}$in from the centre and at right angles from the drilled hole. The bar is drilled $^3$/$_{16}$in at this second mark, again to a depth of 1in.

**4** Four discs, ⅛in thick are cut from the drilled end of the bar either by parting off on the lathe, or by power hacksaw. The discs are marked A, B, C, D in the order in which they are cut from the bar stock. **5** A $^3$/$_{16}$in silver steel rod is inserted in the centre of the discs in the same order as above. Discs A and D are drawn apart from B and C. A piece of $^3$/$_{16}$in silver steel rod, ⅝in long, is inserted in the holes drilled $^5$/$_{16}$in from the centre in discs B and C. The rod, which serves as the displacer crank pin, is silver brazed from the outside of discs B and C. This operation will give a crank throw of ⅝in to the displacer when the con-rod is fitted. **6** Discs A and D are brought close to B and C, so that the gap between A and B, C and D is of just ¼in. A 1½in length of $^3$/$_{16}$in silver steel rod is then inserted in the four discs in the holes drilled $^{25}$/$_{32}$in from the centre. The shaft is silver brazed on the outside of discs A and D, and on the inside of discs B and C. The centre shaft or crankshaft is also silver brazed at this stage on the outside of discs A and D.

**7** With a fine hacksaw the surplus parts of the silver steel rods are cut off. In all eight cuts are required, six for the centre crankshaft between A and B, B and C, C and D and two for the $^{25}$/$_{32}$in throw crank between B and C. All surplus material is removed with a fine file to ensure that there is no binding when the con-rods are assembled.

**A gudgeon block** is prepared from ⅜in brass, ½in long, drilled $^5$/$_{32}$in at one end to a depth of ¼in to take the displacer rod. A set screw, 4 BA, is fitted in the centre of the drilled length. (See Chapter 13 for identical gudgeon block construction). The other end is slotted ⅛in for ¼in depth to take the con-rod. The con-rod is made from bright mild steel bar $^3$/$_{16}$in × ¼in, drilled $^1$/$_{16}$in one end to take the gudgeon pin and brazed at the other end to a big-end. (The method of construction of a big-end is fully covered in Chapter 15.)'

**The twin-conrod system** of the power piston has a con-rod on each side of the displacer rod guide. At one end the con-rods, made of bright mild steel $^3$/$_{16}$in × ¼in, are drilled to take 4 BA bolts. At the other, big-ends are brazed on. In order that the con-rods can function without touching the crank-webs, the rods are bent and shaped at right angles (as in fig 14.6). Careful measurement is essential since the con-rods must avoid touching the engine frame.

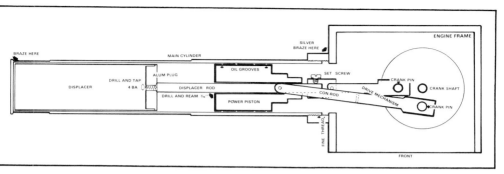

Fig 14.10 *Side elevation.*

## ENGINE ASSEMBLY AND PERFORMANCE

The assembly of 'Sunspot' differs slightly from that of the other engines described in this book. In the first stage the crank assembly is placed at the front end of the slot in the engine frame; the con-rods are inserted through the top cylinder retaining ring, and screwed on to the respective crank pins. The crank assembly is slipped into place and the bearing plates screwed down. The power piston con-rods are next screwed into the power piston collar with a smear of Loctite Screwlock 222 at the tip of the 4 BA or 6 BA retaining bolts.

In the next step the displacer (and displacer rod) are mounted on the power piston and the gudgeon block is tightened by the set screw. The displacer con-rod is placed in the gudgeon block and retained by the gudgeon pin. In the third stage the working cylinder is pushed over the displacer and power piston while a smear of very fine oil is applied to the oil grooves. The cylinder is screwed into its place in the retaining ring. At this stage the crank is turned by hand from the flywheel to ensure that there is no binding. Any adjustments to the displacer stroke are made now.

It is advisable to give a good running-in (without heat) to the engine assembly until the mechanism is working smoothly. This stage of running-in cannot be over-emphasised and should be given its due importance. Finally, before the reflector is mounted, it is advisable to run the engine under power heated by a methylated spirit burner or a not too powerful gas burner (such as a bunsen burner at 50 per cent power). When the engine is working well with minimum heat, the reflector may be introduced. In northern countries, above the 45th parallel (latitude) experiments with the reflector may be best attempted around noon on a bright sunny day. Be careful not to point the reflector in the eyes of the operator or of other people standing by.

The engine frame requires a tilting mechanism to allow it to move from horizontal to vertical positions with minimum effort. This pivot mechanism (fig 14.11) should neither be too low to obstruct the flywheel not too high to make the engine unsteady during its performance. A height of about 1in between the engine frame and the engine base is adequate. If the engine is to be used constantly in a horizontal position, it is worthwhile to devise an engine support so that the main cylinder can avoid undue vibrations.

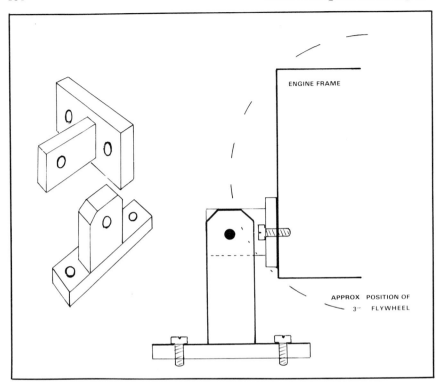

ENGINE FRAME

APPROX   POSITION OF
3"      FLYWHEEL

Fig 14.11 *Tilting mechanism.*

Engine performance depends on the amount of heat generated at the hot
end of the cylinder. With a gas burner or a good methylated spirit flame the
engine easily reached the 1,300 rpm range. With bright sunlight and a good-
sized reflector, speeds of 600-800 rpm can be attained. If the engine is being
used with a reflector only, it is advisable to spray the hot end with a matt
black paint to hasten the absorbtion of heat.

CHAPTER 15

# HOW TO CONSTRUCT 'DYNA', A DEMONSTRATION ENGINE

This engine was built specifically as a demonstration model for exhibitions, lectures and other promotional activities (fig 15.1). It is designed to show that a hot air engine, even in model form, can produce enough power to provide modest work output. The engine can be made to drive a small dynamo (generating enough power to light a bulb), drive a fan to cool the engine radiator and to drive a pump to circulate water between the engine and the radiator. 'Dyna' performs all the above work and that in itself is a great attraction during exhibitions. However the talking point is usually the list near the engine giving details of scrap materials used for the various parts and where they originate. This list is given at the end of the chapter.

Fig 15.1 *Completed engine.*

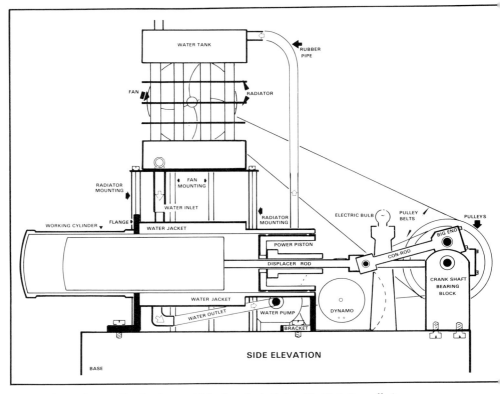

**Above** Fig 15.2 *Engine layout (side elevation).* **Below** Fig 15.3 *Overall view.*

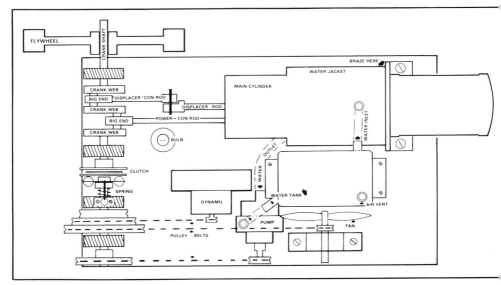

## GENERAL ENGINE LAYOUT

'Dyna' is a single cylinder co-axial engine similar to 'Prova', 'Sunspot' and the Ericsson hot air pumping engine. In planning an engine with ancillary units, the engine layout (fig 15.2) has to incorporate each unit on the engine base without cluttering the area and yet making use of what little space the base provides.

The starting-off point is the engine cylinder — in the author's engine the rare find of a heavy duty Mercedes shock absorber of mono construction and good internal finish was a stroke of luck. An alternative is the construction of a cylinder from a 2in OD mild steel tube with a wall thickness of at least $1/16$in and lapped to a good internal finish for the first $2\frac{1}{2}$in. The base came from surplus electrical equipment but a heavy hardwood base is an excellent alternative.

The crankshaft is fabricated to give the required stroke to the power piston and the displacer. The original crankshaft had three pulleys mounted on it to drive the ancillary equipment but the drive mechanism has now been altered so that the pulley mechanism is a separate unit with a clutch assembly to make the necessary connection between the two. This enables the engine to pick up sufficient speed before the clutch is pushed to activate the pulley drive and perform the work it is designed for.

The cooling mechanism is a complex one but this is for showmanship — there is no need for a pump and a cooling fan blowing through the radiator when a simple thermosyphon system coupled to a water tank will do. The dynamo light gives an impressive effect during demonstrations, especially in a darkened room when the bulb lights the drive mechanism, the clutch assembly and the revolving pulleys.

## METHOD OF CONSTRUCTION

**The engine base** is to a large extent also the engine frame since all the components are bolted directly on to the base. The base came from a large electric component several years old as most of the electrical fittings on it consisted of valves and brass relays. Except for drilling for brackets and supports where necessary no other work was required on the base.

**The main cylinder** is cut and machined from a heavy duty shock absorber. This particular cylinder was of mono construction, that is the walls and bottom are one solid piece without any welding or seam. The material is high quality bright mild steel, with $\frac{1}{8}$in thick walls capable of withstanding great pressure while the interior finish requires no lapping. An alternative cylinder will require different machining procedure.

Work on the cylinder includes the following: **1** Reducing the external cylinder wall thickness by some 50-60 per cent to a length of $2\frac{1}{2}$in from the bottom; **2** Silver brazing or brazing of a flange one-third up from the bottom of the cylinder to enable bolting to the base; **3** Silver brazing of a water jacket to the flange and to the cylinder. This water jacket covers 50 per cent of the length of the cylinder, with inlet and outlet pipes on opposite sides (top and bottom); **4** Fitting of a small bracket at the top end of the cylinder or the water jacket to secure the cylinder to the base.

The water jacket is cut from mild steel tubing such as that used for car silencers and silver brazed to the flange at one end. The other end is reduced by a number of cuts with a fine hacksaw blade and gently hammered inwards to close the opening. A metal filler by Loctite will give adequate sealing almost as good as silver brazing since this side of the water jacket is away from direct heat. This method avoids any possible distortion to the cylinder top due to brazing heat.

The displacer is cut from an outer cover of another shock absorber (some shock absorbers have a sliding cover top which is made from thin gauge mild steel — this is ideal for such work). The tube is brazed at the hot end and then machined in the lathe to reduce further the wall thickness to $^1/_{32}$in, thus reducing the weight of the displacer. An aluminium plug is prepared to seal the open end. This is machined from a dural bar and is drilled and tapped 2 BA to take the displacer rod. An alternative type of displacer may be made from a perfume or spray aluminium container. A modification which includes a method of regeneration, was made after the engine had been working some time — this is discussed at the end of the chapter.

The displacer rod guide is machined from 1in bright mild steel bar. The first ¼in is reduced and threaded ⅜in BSF, a collar ⅜in long is reduced to a diameter of ¾in, and a length of 1¼in is reduced to ⅝in. The guide is centre-drilled, drilled and reamed $^3/_{16}$in, then parted off. The guide is placed in a vice and the ¾in collar is reduced $^1/_{16}$in on one side by filing, to take the power con-rod. A hole is drilled and tapped 4 BA in the centre of the filed area.

The displacer con-rod is machined from $^3/_{16}$in ground silver steel rod, threaded 2 BA at one end to fit into the aluminium plug in the displacer, while the other end fits into the gudgeon block.

The gudgeon block is machined from a ¼in square bright mild steel or brass bar 1in long, drilled $^3/_{16}$in at one end to a depth of ½in, and further prepared for a set screw, (drilled and tapped 4 BA, ¼in from the end). The other end, ½in long is filed down by $^3/_{32}$in to give a recessed appearance (fig 15.3). The recessed block is drilled laterally to take a ⅛in gudgeon pin (or king pin).

The con-rod is fabricated from one piece of bright mild steel flat bar. One end is prepared for a big-end, while the other end is made to fit the gudgeon block. The big-end is constructed in the same manner as the power piston con-rod big end described further on (see fig 15.4).

The power piston is fabricated from a mild steel tube, brazed at one end and machined in the lathe to obtain a good finish. Four oil grooves are cut in the cylinder wall, the front end is faced, centre-drilled, drilled and tapped ⅜in BSF, and finally parted off. The piston is reversed in the chuck, placing it in the jaws with a piece of paper around the diameter to avoid marking. The internal end is faced (with an end mill) to obtain a smooth seating for the displacer rod guide.

The power piston con-rod is prepared in the following manner. A small-end is cut and shaped from a $^3/_{16}$in flat bar, ⅜in × ¼in, drilled in the middle to take a 4 BA bolt and drilled and tapped 4 BA along the narrow edge to take a silver steel rod similarly threaded. A big-end is cut and shaped from a $^3/_{16}$in flat bar, 1¼in × 1in, drilled in the middle to take a ⅝in bush or ball bearing. One narrow (1in) edge is also drilled and tapped 4 BA to take a silver steel rod threaded 4 BA. The big-end is drilled and tapped to take two 8 BA bolts, one on each side of the bearing. The big-end is then sawn in half with a fine hacksaw blade. To facilitate bolting the two halves together, the outer half of the big-end 8 BA holes are redrilled ³/₃₂in. The big-end assembly is tightly bolted together and the ⅝in hole is redrilled to enlarge the reduced bore due to the hacksaw blade cut.

Thirdly, a length of ³/₁₆in silver steel rod is threaded 4 BA at both ends to a length of ½in — one end to fit the big-end, while the other end is threaded into the small end. Adjustments to the length of the power piston con-rod are made by screwing or unscrewing either end. A lock-nut at each end ensures regidity in the con-rod assembly.

**The drive mechanism** is fabricated from a mild steel flat bar ³/₁₆in thick and from ³/₁₆in silver steel rods. Three webs are cut, 1¼in × 1in, drilled and reamed ³/₁₆in. The holes are marked in a right angle formation with the distances between centres shown in fig 15.5. The webs are then drilled and reamed together and the crankshaft assembled (fig 15.6). The completed crank assembly has a single ¾in throw for the power piston and a 1³/₁₆in throw for the displacer, with a 1½in shaft end for the flywheel and a 1in length for the clutch disc. The webs, crankpins and crankshaft are either silver brazed or fitted together with steel pins (see the 'Sturdy' fabricated crankshaft).

The crankshaft (fig 15.6) is mounted on two main bearing blocks made of

**Right** Fig 15.4 *Big-end construction.*

**Below** Fig 15.6 *Crank-web marking.*

**Below right** Fig 15.6 *Fabricated crankshaft assembly.*

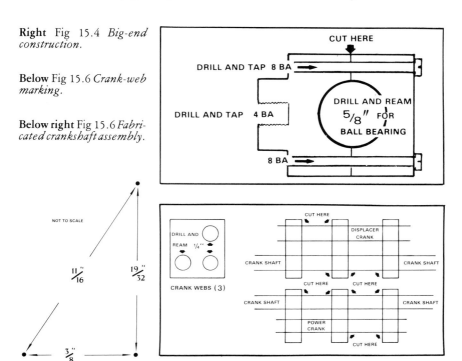

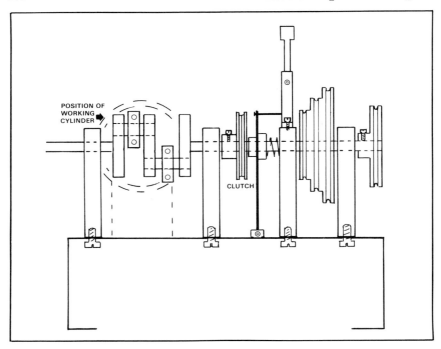

POSITION OF
WORKING
CYLINDER

CLUTCH

**Above** Fig 15.7 *Front elevation.*

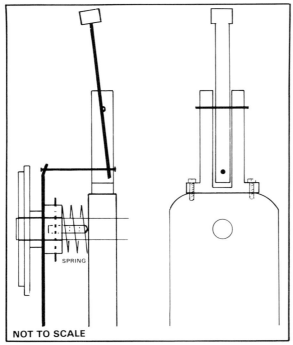

SPRING

NOT TO SCALE

**Left** Fig 15.8 *Clutch assembly.*

⅜ in bright mild steel with brass bushes as bearings. The blocks are fitted to the base after the position of the working cylinder has been fixed. Any slight adjustment of the cranks in relation to the power and displacer con-rods should be done at this stage of construction. The big-ends are made to run with brass bushes as bearings or on ball bearings. Brass bushes may be inserted in halves in the big ends during assembly, whereas ball bearings, ID ³/₁₆in, OD ⅝in (or alternative OD) are inserted in the crankshaft in the final stages of fabrication. The clutch plate is fitted with a collar and set screw secured to the inner crankshaft end. The engine should be run-in and tried out at this stage.

**The pulleys and clutch** are one assembly on a single drive shaft, mounted on two main bearing blocks. A general idea of the layout of this assembly may be obtained from fig 15.7 seen together with fig 15.3. The outer pulley on the left is the pump drive. The centre multiple pulley block drives the fan and the dynamo. The pulleys are secured to the shaft by means of set screws. The clutch assembly (fig 15.8) while seemingly complicated, is in fact a delightful mechanism and is quite efficient in operation. Reference to fig 15.7 is suggested before, and while, the machining procedure is in operation.

The shaft has a ³/₃₂in slot, drilled and filed to a length of ½in, ¼in from the clutch end. The clutch plate is fitted to a two-step collar, ⁵/₁₆in on the inside (against the plate), ½in on the outside. The outer part of the collar is drilled ¹/₁₆in through to take a steel pin. A ¾in wide (¹/₁₆in thick) brass lever, drilled ⅜in and with facility to pivot at the base, is mounted between the clutch plate and the stepped collar. This lever is made to pull back the plate when required; it is made to pivot by soldering at the bottom end a ¹/₁₆in brass pin 1¼in long, mounted with 2 BA cheese-head bolts drilled ¹/₁₆in and secured from or in the base. The top of the lever has a ⁵/₆₄in hole drilled ½in above the collar — this takes the connecting rod to the clutch drive lever situated on top of the main bearing block.

The clutch-drive lever and bracket assembly are constructed from brass and dural respectively. The lever is made from a brass strip, ¼in wide, ⅛in thick and about 2in long; a ¹/₃₂in hole is drilled laterally in the centre and a ⁵/₆₄in hole near the bottom through which is inserted the clutch plate connecting rod consisting of a long 10 BA bolt with three nuts. The bracket is cut and shaped from ¼in dural (fig 15.8), and drilled to take the clutch drive lever. The bracket is bolted to the bearing block.

**The assembly of the clutch** is as follows. 1 The lower lever is inserted in the collar between the plate and the wide collar; 2 The clutch plate is secured to the collar by bonding or soldering; 3 A ¼in compression soft spring ½in long is inserted between the collar and the main bearing block (with a fine washer between the spring and the block). The action of the spring is to keep the two plates against each other — disengagement is by pulling back one plate against the spring; 4 A steel pin is inserted in the collar through the slot in the pulley shaft ensuring that the pin does not slip out; 5 The lower lever pivot arrangement is fitted to the base while the main bearing clock is bolted on the base. At this stage the clutch assembly is tested. The lower lever should pull back the clutch plate against the compression spring. On being released the clutch plate should return to face against the clutch plate on the crankshaft.

Any alignment required should be done at this stage; **6** The upper clutch-drive lever mechanism is assembled and bolted on the main bearing block. The lever action is such that when the lever top knob is pushed in the direction of the drive mechanism, it pulls back the lower clutch lever against the spring and disengages the clutch plates. The connecting rod, a 10 BA bolt 1in long, is inserted in the top hole in the lower clutch lever and a nut is screwed on permitting slight movement of the connecting rod. A smear of Loctite Screwlock 222 ensures that the nut remains in place. The other end of the bolt is inserted into the upper clutch lever with a nut on either side held in place with Loctite. The whole assembly is checked for efficiency and smooth action. The upper lever action should be stiff enough to counter the action of the spring. Finally two washer-type rubber rings are bonded on to the clutch plates. These are best cut from car tyre rubber and fixed with Loctite super glue with the rough side out, ie, rough sides facing each other. The pulleys are screwed on the drive shaft and the remaining main bearing block is bolted to the base. The whole assembly is checked for alignment.

The general idea of this clutch assembly is that the plate on the pulley side can turn the shaft while at the same time sliding sideways to engage or disengage. There are other types of clutches, such as the cone clutch, that can be used on this engine.

**The radiator** can be contructed from two brass or tin containers (tanks) with a number of connecting ¼in brass tubes. Three or more layers of ¹/₃₂in brass sheets are inserted between the two tanks to serve as heat dissipaters and to give a better image of a radiator. The top and bottom covers of the tanks as well as the brass sheets are drilled together to obtain correct alignment of the brass tubes. The tubes are soft soldered to the tanks ensuring that the joints are water tight. The brass sheets are placed at equal distances along the height of the radiator. The filling tube, inlet and outlet bends are soldered on and then the whole radiator assembly is mounted on four pillars made of hexagonal brass or silver steel rods to a height about 1½in to 2in above the working cylinder (fig 15.10).

**The water pump** The casing of this particular pump (fig 15.9) was made from a solid cube of perspex 1½in × 1½in × 1¼in, which was drilled through to take a ⅛in shaft. The hole was then bored half-way through its depth to an internal diameter of ½in and finished off with a boring bar to give a 1in bore with a flat sided bottom. A ¼in perspex cover with a ⅜in boss was prepared and drilled with a ¼in hole to serve as a water inlet. The cover was fitted to the pump body by means of four 6 BA screws. (A gasket of thin rubber material may be required to give good sealing).

When the actual position of the pump and the rotation of the engine have been determined, the water outlet on the pump body is marked and drilled. If the impeller is turning clockwise, the outlet hole is on the top left corner of the pump body; if anticlockwise, the outlet is on the top right corner. The impeller may be constructed from perspex or may be cut out from a solid rubber block with a sharp knife, the important point being that the impeller should have several equal blades. The shaft must be tight fitting in its housing with the impeller sitting close to the pump body. 'O' ring seals may be used to control any water seepage through the shaft. A small pulley is

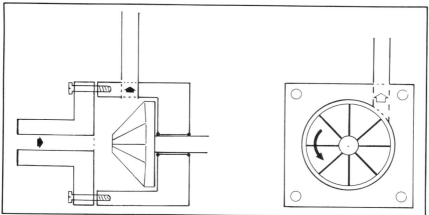

**Above** Fig 15.9 *Water pump.*

**Right** Fig 15.10 *Ancillary equipment.*

fitted to the shaft. Since fast rotation helps in giving a good flow, the pulleys should have a 1:6 ratio. The pump is bolted to the base with its pulley in line with the pulley on the drive shaft. Water pipe connections are made to the underside of the working cylinder (inlet to the pump) and to the top radiator bend (inlet to the radiator).

**The dynamo** installation is a matter of experimentation to find a small motor that gives sufficient current to light a small electric bulb. The bulb used on this engine was 6V.5A; there is sufficient power for one to light brightly and for two to light with less intensity. It is recommended that one bulb is connected intially and placed on a lamp-post type fitting near the drive mechanism; the pulleys should have at least an 8:1 ratio.

**The fan assembly** is a hotch-potch of materials which together make another impressive piece of equiment. In actual fact if the revolutions are sufficiently high quite a breeze is created and as soon as the radiator starts to warm up the cooling effect is quite astonishing. The fan blades are cut from $1/16$in aluminium sheet, four or more blades being marked and cut ensuring that the size and pitch of each blade are identical. The hub of the fan is bonded to a small pulley fixed on a shaft fitted in a brass tube in a dural block, the whole assembly standing on two pillars directly in line with the centre of the radiator (fig 16.9). A cylinder sleeve mounted on the radiator surrounding the fan ensures that the breeze is contained on the radiator with greater efficiency. The various belts are cut from a 1.6mm Loctite nitrite rubber 'O' ring belting, the ends of each belt being joined by Loctite Superglue 3.

## PERFORMANCE AND MODIFICATIONS

The author has made a modification to the basic engine design which may be of interest to modellers. The original displacer made out of mild steel was considered to be slightly heavy and an alternative displacer was used. The displacer in use is a deodorant aluminium container. This has an OD of 1 $37/64$in leaving an annular gap of $11/64$in on the diameter and $5/64$in or just over $1/16$in on the radius of the cylinder. A spring wound tightly from $1/16$in steel wire is inserted in the cylinder along the length of the displacer and its stroke. The net result of this modification is twofold; a quicker response to heating and longer run-on with the burner turned off. The power of the engine is not affected.

A second modification is the incorporation of a 'snifter valve' in the power piston. Originally a hole was drilled in the crown of the power piston between the inner wall and the displacer rod guide bush to facilitate the running-in stage (see Chapter 8). The hole was drilled and tapped 0 BA and at first a plug bolt was used to seal the hole. As an experiment, the plug bolt assembly was modified and replaced by a snifter valve which operates when the mean pressure inside the engine falls below atmospheric. The construction of this valve is a simple matter requiring patient bench fitting and lathe work.

A length of ¼ in OD brass bar is reduced and threaded 0 BA at one end, then drilled through $3/64$in to take a long 10 BA bolt, long enough to exceed the length of the brass bar by ¾ in (such bolts are usually used on the older type of electrical component). A fine compression spring ¼ in long is also required to fit the 10 BA bolt almost exactly. (Such springs are used under press keys of miniature keyboard musical instruments.) The seating of the bolt is the part of the assembly which requires greatest attention. The bolt should preferably have a 'cheese' head as this has a better seating capability. The use of a fine rubber washer is recommended. The snifter valve is assembled with the bolt inserted from the 'inside' end, ie, opening into the power cylinder, the spring mounted externally and a nut screwed on allowing movement to the valve when feather light pressure is applied. The valve is then screwed into the piston crown with a smear of Loctite Screwlock 222.

## Materials used in the construction of 'Dyna'

| | |
|---|---|
| Engine base | Surplus radio equipment base |
| Main cylinder | Mercedes heavy-duty shock absorber |
| Water jacket | Scrap silencer tube |
| Displacer | Yaxa deodorant container |
| Power piston | Cast iron water conduit pipe |
| Crank webs, big ends, main bearing blocks | Scrap bright mild steel offcuts |
| Radiator | Two old Oxo tins, brass sheets from scrap electrical equipment (brass tube purchased) |
| Fan | Scrap aluminium sheet and brass material |
| Radiator and fan supports | Scrap hexagonal pillars used in electrical equipment |
| Flywheel | Old German sewing machine (aluminium casting was also made and used) |
| Dynamo | ex-Pye cassette tape recorder drive motor. |

## Engine specifications

| | |
|---|---|
| Base | 11in × 6½in |
| Working cylinder | 2in OD |
| | 1¾in ID |
| | 8in internal length |
| Power piston | 1¾in OD |
| | 1½in length |
| Guide bush | 1in OD unmachined |
| | 1¾in length |
| | $3/16$in ID |
| Displacer | first — 1⅝in OD |
| | second — 1 $37/64$in OD |
| | 4½in length |
| Power stroke | ¾in |
| Displacer stroke | 1 $3/16$in |

CHAPTER 16

# HOW TO MEASURE ENGINE PERFORMANCE (SPEED, TORQUE AND POWER)

No model engineer is satisfied with his efforts unless he finds out the actual performance of his engine. The two questions that come to mind are how fast? and what power? Measuring the revolutions of a hot air engine is no easy matter, as every modeller finds out sooner or later. The normal tachometer pushed against the shaft stops an engine dead. After all we are dealing with models of engines whose shaft power, even in the larger type of hot air engine, is always relatively low. In order, therefore, to obtain an accurate reading of the revolutions the modeller has to resort to a mechanical revolution counter which puts little or no friction on the engine, or to an electronic instrument which measures speed without any contact with the engine. There are a number of different revolution counters which are suitable and reliable. The one described here is one of the more accurate and easier to construct.

## MECHANICAL REVOLUTION COUNTER

This counter consists of a worm and wheel assembly which when coupled to the engine output shaft causes the wheel to turn without effort. A stop watch is used to time a given number of turns. The construction is simple but precision is essential to ensure minimal friction.

A bracket is contructed from a brass strip ⅛ in thick and ¼ in wide, with space for the worm gear to turn without binding. Two holes are drilled at the same height in the lateral arms of the U-bend. The size of the holes depend on the shaft used on the worm gear. If a ⅛ in shaft is used, a slightly smaller hole is drilled first and then a ⅛ in hole is drilled from side to side. The holes are finished off with a reamer. Two holes are drilled in the bottom of the bracket to take 4 BA screws. The gear wheel is mounted on a dural base ¼ in thick. A shaft to fit the wheel is screwed into the base. The wheel is then placed through the shaft with a couple of washers underneath, and the assembly is secured from above by a collar. The wheel should be free enough to revolve easily with the merest finger flick.

In the third stage of construction the U-bracket with the worm attachment is placed along the wheel and the securing holes marked for drilling. The

**Above** Fig 16.1 *Revolution counters. The revolution counter on the left was constructed from parts of a junior engineering set while the one on the right is described in the text. Putting together the assemblies was relatively easy.*

**Right** Fig 16.2 *Test rig and accessories on engine.*

holes are drilled slightly larger than the bolt diameter to allow for lateral adjustment of the bracket. Two important points should be kept in mind: The centre height of the worm should be that of the wheel and secondly the worm thread should just touch the teeth of the wheel. Finally the drive shaft of the worm is fitted with an adaptor or a universal drive which also fits the engine drive shaft. A simple and fairly efficient drive may be made from a short piece of soft rubber tubing joining the two shafts (fig 16.2).

One way to test the efficiency of this instrument is to insert a piece of cardboard in a slot in the adaptor and to blow on one side of the cardboard. A soft blow should be sufficient to turn the worm and wheel a few turns. Another revolution counter more suitable for the smaller engine (up to 10cc) can be constructed from parts of an electric clock which has seen better days. These clocks usually have one or more worm and wheel sets. Greater precision is required in fitting together this counter due to the size of gears and shafts.

The revolution counter should be mounted on a stand with a heavy base. In constructing this stand allowance should be made for the counter to be adjusted for height. A typical stand can be made from a 1 ½ in OD iron bar by 1in high into which a ¼ in steel rod is inserted and bonded vertically. A collar with a grub screw supports the revolution counter base which is drilled to fit the steel rod. A grub screw fitted laterally in the base makes the assembly quite rigid.

In using the revolution counter the wheel and the base are marked with pointers to show the starting mark. The counter is placed against the engine shaft so that the shafts are exactly aligned and the universal drive fitted. The engine is run and when the speed to be calculated is reached, just as the pointers coincide, the stop watch is started. The watch is stopped when a predetermined number of turns of the wheel have been made. If the wheel has 95 teeth and has gone round 5 times in 20 seconds, the speed works out as follows:

$$\frac{95 \text{ teeth} \times 5 \text{ turns}}{20 \text{ seconds}} \times 60 = \text{rpm ie,} \frac{475}{20} = 23.75 \times 60 = 1,425 \text{ rpm}$$

## ELECTRONIC REVOLUTION COUNTERS

Electronic revolution counters of the types described below have two distinct advantages over the mechanical types. The equipment does not touch the engine or create friction and setting up the measuring equipment takes very little time.

**The stroboscope** has a probe or a wand with a flickering light placed near the moving mechanical drive (fig 16.3). Unfortunately this has to be done in a darkened room for the flickering bulb in the probe to throw sufficient light on the moving parts. The method of operation is fairly simple — the engine is run, the equipment switched on and the probe placed near the mechanical drive (the connecting rods are the easiest 'readable' target). The control knob on the dial is shifted slowly until the moving parts appear stationary in the light of the probe. A reading is taken on a calibrated dial and the speed read off. Usually the pulses are set in cycles per second and any reading is therefore calibrated either in pulses and multiplied by 60, or the dial calibration is already translated into actual speed. Stroboscope components are listed in Appendix 2.

**The digital display tachometer** fig 16.5 (a & b) utilises an infra red source and sensor. The main requirement in operating the tachometer is to fit a piece of stiff cardboard in the shape of a long and narrow tongue bonded to the

**Above** Fig 16.3 *Strobo-scope.*

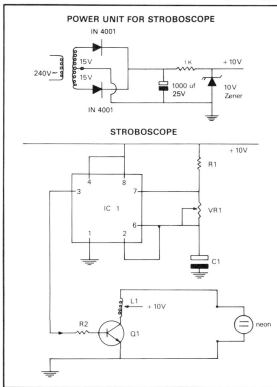

**POWER UNIT FOR STROBOSCOPE**

**STROBOSCOPE**

**Right** Fig 16.4 *Stroboscope electrical circuit* (see also Appendix 2).

**Above** Fig 16.5 *Tachometer (open).*

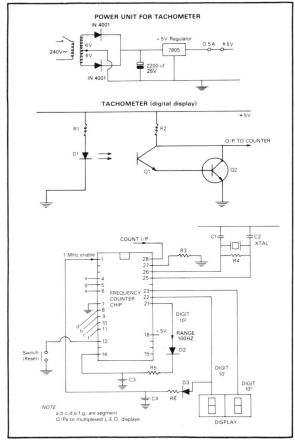

**POWER UNIT FOR TACHOMETER**

IN 4001

240V~  6V
6V

+ 5V Regulator
7805

0.5A  +5V

2200 uf
25V

IN 4001

**TACHOMETER (digital display)**

+5V

R1

R2

O/P TO COUNTER

D1

Q1

Q2

COUNT I/P

C1    C2
XTAL

1 MHz enable    1

28    R3
27
26
25    R4

e    4
g    5
a    6

FREQUENCY    23
COUNTER    22
CHIP    21

7
8
9
10
11

d
b
c    f

18    +5V

DIGIT
10²

RANGE
100HZ

Switch
(Reset)

12

14    15

D2

DIGIT
10¹

R5

C3

D3

DIGIT
10⁰

C4    R6

DISPLAY

*NOTE*
a,b,c,d,e,f,g, are segment
O/Ps to multiplexed L.E.D. displays

**Left** Fig 16.6 *Tachometer electrical circuit* (see also Appendix 3).

flywheel. During the revolutions the cardboard tongue slips between the infra red source and sensor serving as an interruptor. The source and sensor are fitted in a plastic holder and mounted on a stand which is placed adjacent to the flywheel. The instrument itself may be placed nearby where it can be read with ease. The digital display tachometer has two advantages over the stroboscope in that it can be operated in full light and there is no probe to hold. Components of this instrument are listed in Appendix 3.

The digital display gives a reading in cycles per second. Therefore an 07 on the display means $7 \times 60 = 420$ rpm. A display of 12 means $12 \times 60 = 960$ rpm. The two figure display can show up to 99 ($\times 60 = 5,940$ rpm) which is higher than a model engine is likely to go. However with minor modifications to the circuit the display can be made read up to 4 figures, ie, 9,999 cycles per second ($\times 60 = 599,940$ rpm).

## HOW TO MEASURE TORQUE

The power generated by an engine is measured in watts or horsepower, (746W = 1hp). Power cannot, however, be measured directly as it is the product of two other engine parameters: speed (rpm) and torque, both of which can be measured directly at the engine crankshaft. The torque of an engine is the turning or twisting force it exerts at its shaft — the torque naturally varies with the engine speed.

To understand how torque is measured, one must imagine that the twisting force at the engine shaft (fig 16.7a) is equivalent to a force W acting at a radius X (fig 16.7b).

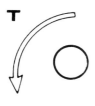

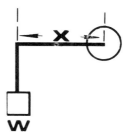

Fig 16.7(a) *Torque.*                    Fig 16.7(b) *Weight × distance.*

The torque is then equal to the product WX. In the SI system of units, W is measured in newtons and X in metres, so that the SI unit of torque is the newton-metre (Nm). For model engines this unit is much too large, so we express W in grams and X in centimetres, and we measure torque in gram-centimetres (gcm). This necessitates the use of a conversion factor when calculating power.

When an engine is running off load, ie, nothing attached to the shaft, all the power generated is used to overcome internal friction. At this speed, the output power and hence the torque available at the shaft is zero. If a load, (eg, a generator) is attached to the engine shaft, the speed of the engine decreases. Some of the power which was previously being wasted to overcome friction is now being delivered in the form of shaft power. At this stage the torque of the engine is exactly balanced by the equal and opposite torque exerted by the load. To measure the torque at this particular speed, we must remove the load (ie, generator) and attach instead a torque measuring device, which will then be adjusted until the engine speed returns to its original level, that is at the generator load level. This torque measuring device is called the 'torque testing rig', and its principle can be unserstood from fig 16.7(a) and (b). The torque testing rig exerts a counter-torque by means of weight $W_1$ hanging from radius $X_1$. This counter-torque $W_1X_1$ balances exactly the engine torque WX when the engine is running at a steady speed. If $W_1$ is in grams and $X_1$ is in centimetres, the product $W_1X_1$ is the engine torque in gram-centimetres.

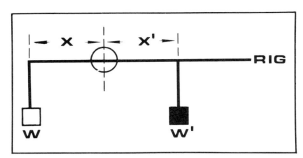

**Left** Fig 16.7(c) *Counter-torque diagram.*

**Below** Fig 16.7(d) *Graph.*

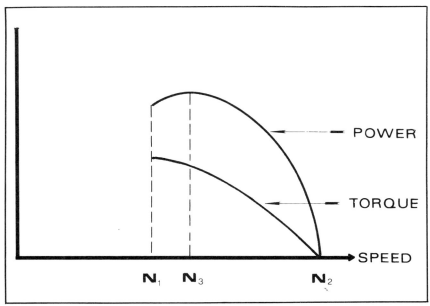

A typical test rig with its accessories is seen in fig 16.8, while a rig assembled for engine testing is seen in fig 16.2. The torque test rig consists of a brass drum which slides onto the engine shaft and is tightened by a grub screw. The drum rotates freely inside a PTFE, nylon or wooden friction clutch. The clutch is made in two halves. Two set screws can be used to tighten the two halves onto the drum to increase the friction. The clutch has an arm attached to one side and an adjustable counter-weight screwed on the other side. The arm is calibrated in distance (centimetres or inches) from the centre point of the drum. The torque test rig is mounted on the engine shaft and the drum tightened. The revolution counter is then fitted on to the engine shaft (see fig 16.2).

To initiate the test a suitable small weight is placed on the arm, (eg, 10 grams at a distance of 10cm), the clutch is loosened and the engine is started with the arm pointing downwards. Once the engine picks up speed, the clutch set screws are tightened. This causes an increase in frictional torque on the engine shaft and the engine speed to decrease. The screws are adjusted until the test rig is observed to fluctuate about a mean horizontal position. At this stage the engine rpm corresponding to the given torque (which is 10gm × 10cm = 100gcm) can be read off the revolution counter. If for example, a speed of 600 rpm is observed, then the engine power:

$$W = \text{grams} \times \text{centimetres} \times \text{rpm} \times .00001026$$
$$W = 10\text{gm} \times 10\text{cm} \times 600 \text{ rpm} \times .00001026 = 0.6 \text{ watts.}$$

This procedure can be repeated for a range of different torques, (the torque being varied by either changing the mass or altering the radius-arm length). For each torque value, the engine speed is read and the power calculated. Three engine speeds of particular interest are $N_1$, $N_2$, and $N_3$ (fig 16.7d). $N_1$ is the engine stalling speed, that is the lowest speed at which the engine can run. Any further increase in torque at this point will cause the engine to stop. $N_2$ is the no-load speed, and is the maximum speed at which the engine can run (since torque is zero at this point, the output power is also zero). Somewhere in between lies $N_3$, the speed at which the engine delivers maximum power. This speed is one of the most important parameters of any hot air engine.

A simple torque rig can be constructed in the following manner. Required are: A brass drum, a hardwood or PTFE block, an aluminium strip 6in long × ½in wide × $^{1}/_{16}$in thick, two 4 BA studs; two fine compression springs to fit the studs. The size of the drum, although not critical, should in practice be about ¾in diameter. Machining the drum consists of drilling a bore to fit the engine crankshaft, stepping or recessing the drum and reducing the drum on both sides to obtain two collars, which are tapped to take grub screws. The hardwood or PTFE block is of the same thickness as the inner grooved part of the drum and is bored to fit the same diameter of the grooved drum. A fine saw cut is made across the diameter of the bore while the top corners of the block are sawn off. The top and bottom halves are drilled and the bottom half tapped to take the studs. The aluminium strip is bonded or rivetted to the bottom part of the block so that the strip is perfectly in line with the bottom half. The arm is marked in cm/in distance from the mid-centre point of the

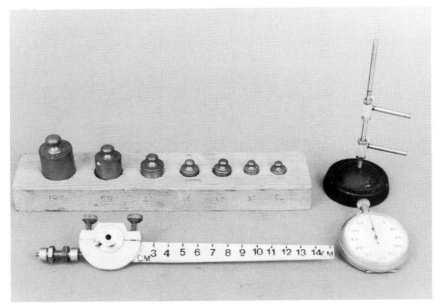

Fig 16.8 *Test rig and accessories.*

drum bore in the block. The calibration, apart from the marking, is scored by notches on the arm.

It will be noticed that on assembly the weight of the arm will pull the arm downwards towards the perpendicular. Therefore a counterweight must be fitted on the side of the block opposite the arm to balance, so that when the instrument is mounted it remains perfectly horizontal, and when rotated by hand on the engine shaft it will regain this position repeatedly. There are several methods of fixing the counterweight. One is to insert a stud and add on nuts or other weights until a perfect balance is achieved. The rig is assembled with the drum into the block, the studs in place, the springs on the studs above the top block and the nuts to retain and tighten the springs on the studs. The rig is now ready for the first test run. A simple device such as a wooden stand with an upright beam to which are fixed two dowels, placed near the end of the arm, will prevent the arm from excessive movement and allows fine adjustment to the calibration when friction is applied.

APPENDIX 1

# MATERIALS USED IN MODEL ENGINE CONSTRUCTION

Screws, bolts, nuts, washers, spring washers.

Silver steel or spindle steel roads — high quality precision for displacer rods and crankshafts. ⅛in, $3/16$in, ¼in mostly used.

Brass round bars — ¼in, ⅜in, ½in mostly used for bushes, bearings and guides.

Dural flat bars — different thickness, $3/16$in to ½in for flanges, cylinder plates, engine bases etc.

Dural round bars — different diameters, off cuts from ¼in to 2in, the latter for cooling fins.

Cast iron bars — off-cuts for making power pistons — around 1in.

Tubes — stainless steel, bright mild steel etc — ¾in to 1in OD for cylinder work.

Shims — brass, for packing; steel, for regenerators if being used.

Brass tubes (as used for propellor shafts) — for small bushes and guides.

PTFE, Teflon, Rulon — off-cuts for to use for bushes, sealing rings etc.

Gasket material — various thicknesses up to $1/16$in.

Other consumable items:

Very light fine lubricating oil.

Loctite glues — nutlock, stud lock, screwlock.

Bonding agents; Loctite, Araldite, Super Epoxy etc.

Light grease.

Cutting oil.

Very fine grinding paste.

Polishing compound.

APPENDIX 2

# STROBOSCOPE COMPONENTS LIST

(See also fig 16.4)

| | |
|---|---|
| $R_1$ | 10K 0.5W oxide resistor |
| $R_2$ | 1K 0.5W oxide resistor |
| $Q_1$ | BC 107 transistor |
| $IC_1$ | 555 timer |
| $VR_1$ | 100K variable potentiometer |
| $C_1$ | 2 microfarad electrolytic capacity |
| $L_1$ | Centre tapped transformer by selection |
| | Miniature neon |

APPENDIX 3

# TACHOMETER DISPLAY COMPONENTS LIST
(See also fig 16.6)

| | |
|---|---|
| $R_1$ | 270 ohm 0.5W oxide resistor |
| $R_2$ | 1.0K ohm 0.5W oxide resistor |
| $R_3$ | 100K ohm 0.5W oxide resistor |
| $R_4$ | 22 Meg ohm 0.5W carbon resistor |
| $R_5$ | 10K ohm 0.5W oxide resistor |
| $R_6$ | 10K ohm 0.5W oxide resistor |
| $C_1$ | 47 Picofarad polystyrene capacitor |
| $C_2$ | 47 Picofarad polystyrene capacitor |
| $C_3$ | 68 Picofarad polystyrene capacitor |
| $C_4$ | 68 Picofarad polystyrene capacitor |
| $D_1$ | Infra-red source (RS 306 77) |
| $D_2$ | IN 4148 |
| $D_3$ | IN 4148 |
| $Q_1$ | Infra-red sensor |
| $Q_2$ | BC 107 |
| Display | RS 2-digit 0.5in multiplexed display |
| Frequency | RS 7216 C (part no 306 837) |
| Crystal | 1MHz (RS 307 761) |
| Switch | Push-to-make |

Note: RS = Radio Spares

# BIBLIOGRAPHY

*Stirling Engines,* by Professor Graham Walker, published by Clarendon Press, Oxford. Second edition 1980

*Stirling-cycle engines,* by Andy Ross, published by Solar Engines Inc, Phoenix, Arizona, USA Second edition — 1981

*Liquid piston Stirling engines,* by Dr Colin D. West, Ph.D, of Oak Ridge National Laboratory, published by Van Nostrand Reinhold Co of New York — 1983

In addition *Model Engineer* published by Model and Allied Publications Ltd (MAP) and *Engineering in Miniature* published by Tee Publishing, often print contributions by enthusiasts of Stirling engines and other hot air engines. These are all worth reading since they are a great source of knowledge and specific information pooled and exchanged by experts in this field.

One should mention the three articles which appeared in *Philips Technical Review:*

1947, Volume 9, No 4; pages 97-104 — *Fundamentals for the development of the Philips air engine,* by H. de Brey, H. Rinia and F. L. van Weenan.

1947, Volume 9, No 5; pages 125-134 — *The construction of the Philips air engine,* by F. L. van Weenan.

1959, Volume 20, No 9; pages 245-262 — *The Philips hot gas engine with rhombic drive mechanism,* by R. J. Mejjer

and a number of articles which appeared in the last century in *The Engineer.* However one particular study is a 'must' for modellers — the article by T. Finkelstein which appeared in *The Engineer* in 1959:

Volume 207, pages 492-497, 522-527, 720-723; *Air Engines* by T. Finkelstein.

# INDEX

# *Other books for enthusiasts!*

PSL Model Engineering Guide:
## Introducing the Lathe
Stan Bray

The lathe is one of the most essential tools in any home workshop, and this book not only describes its many different functions but also explains how to choose the right lathe for your own purposes and how to use it safely. Packed with photographs and clear diagrams, it is a must for anyone considering taking up model engineering as a hobby, but also includes much useful information for 'old hands'.

*96 pages, illustrated, paperback.*

PSL Model Engineering Guide:
## Introducing Benchwork
Stan Bray

Beginning with an examination of the workshop bench itself and how to build one, this vital book for all model engineers explains marking up from working drawings as well as a myriad of operations such as threading, drilling, welding, brazing and soldering, bending and polishing, showing which tools are used for each job and how they should be handled to prevent accidents.

*160 pages, illustrated, paperback.*

## PSL Complete Guide to Model Railways
Michael Andress

In over five hundred pages, this unrivalled compendium covers basic trackwork and electrification, scenic modelling and structures, the operation of a layout and a host of other vital subjects. There are sections on track schemes and railway types, choice of models and materials, branch line and narrow-gauge railways, modern railways and a great deal more.

*516 pages, illustrated, hardback.*

These and the wide range of other Patrick Stephens books on subjects as diverse as astronomy and sailing or military history and motor cycles, can all be obtained through your local bookshop. For more information, please send for a free copy of our complete illustrated catalogue of new books and books in print to: **Patrick Stephens Limited,** Denington Estate, Wellingborough, Northants, NN8 2QD (*Telephone 0933 72700*).